Development of the Craniofacial-Oral-Laryngeal Anatomy

A Review

Development of the Craniofacial-Oral-Laryngeal Anatomy

Ray D. Kent, Ph.D.
Houri K. Vorperian, M.A.

Waisman Center on Mental Retardation and Human Development
University of Wisconsin–Madison

SINGULAR PUBLISHING GROUP, INC.
SAN DIEGO · LONDON

Singular Publishing Group, Inc.
4284 41st Street
San Diego, California 92105-1197

Typeset in 10/12 Century Schoolbook by CFW Graphics
Printed in the United States of America by McNaughton and Gunn

Library of Congress Cataloging-in-Publication Data

Kent, Raymond D.
Development of the craniofacial-oral-laryngeal anatomy : a review / Ray D. Kent and Houri K. Vorperian.
p. cm.
Includes bibliographic references and index.
ISBN 1-56593-425-3
1. Skull—Growth. 2. Face—Growth. 3. Mouth—Growth Larynx--Growth. I. Vorperian, Houri K. II. Title.
[DNLM: 1. Face—anatomy & histology. 2. Skull—anatomy & histology. 3. Skull—growth & development. 4. Larynx—anatomy & histology. 5. Larynx—growth & development. 6. Speech—physiology. WE 705 K37d 1995]
QM535.K46 1995
611′.92—dc20
DNLM/DLC
for Library of Congress

95-31082
CIP

Contents

Preface

Speech is produced with an anatomic system that serves a number of functions besides speech and that changes its size and shape throughout the human lifespan. The rates of change vary considerably across age periods and across the structures that comprise the speech production system. The periods of infancy and pubescence are associated with especially remarkable rates of growth, but other periods are by no means static. There is evidence to support the idea that the speech production system as a whole undergoes continual changes from its embryological state to its terminal state in old age. A general issue in understanding speech development is to account for the growth of the speech production system and the implications of this growth for studies of articulatory-acoustic relations, speech motor skill acquisition, phonologic development, and the effects of craniofacial anomalies and various pathologies on speech development in children.

Growth includes changes in size and possibly shape of the individual components of a given system. These changes can be considered at the levels of both micro- and macroanatomy. In addition, structures may change in their relative positions, such that the entire configuration has a developmental profile. This appears to be the case with the human vocal tract, which assumes various developmental configurations. The purpose of this monograph is to summarize the growth patterns of the human craniofacial, oral, and laryngeal anatomy from birth to adulthood. A specific objective based on this anatomic description is to identify structural changes that relate to the ontogeny of the system for speech production, mastication, and swallowing. The major questions to be addressed are as follows:

1. To what degree do growth patterns differ for the various parts of this complex system?
2. To what degree are changes in size accompanied by changes in shape, and what is the nature of shape changes for individual structures?
3. How does re-formation of the overall anatomy relate to major events in the development of speech and other oral motor behaviors?
4. What variables determine the rates and patterns of growth of the various components of the craniofacial-oral-laryngeal system?

This review emphasizes size and shape changes in the craniofacial skeleton, nasopharynx, lips, tongue, and larynx. Selected information on histology is presented, but the review is concerned mainly with macroanatomy. Notes on clinical applications and summary tables of selected major sources on anatomic development are also included.

We undertook this review as a kind of introductory piece to a more general examination of speech development in children. Anatomy is only part of the story of how speech develops, but it is an important part. Because we did not find a comprehensive summary of the developmental anatomy of the structures that constitute the vocal tract in the literature, we decided that it would be a worthwhile effort to review the available information and attempt to synthesize it into a coherent picture of the development of the speech production system. The product of this effort might be subtitled, "How to Grow a Vocal Tract."

Acknowledgments

This research was supported by NIH research grants DC00319 ("Intelligibility Studies of Dysarthria") and DC0082 ("X-ray Microbeam Speech Production Database") from the National Institute on Deafness and Other Communication Disorders, National Institutes of Health. We are grateful to Fred Gruber, Megan Hodge, and Joel Kahane for drawing our attention to helpful sources of information. David Kuehn made a number of helpful comments on a draft of the manuscript. We also thank Michael Laier for his skill in the preparation of the growth charts and Seta K. Wehbe for her meticulous translations of the French articles. We appreciate the support of Leonard L. LaPointe, Editor of the *Journal of Medical Speech-Language Pathology,* and the fine people at Singular Publishing Group, who brought our work to a finished product.

Introduction

This review summarizes the growth patterns of the human craniofacial, oral, and laryngeal anatomy from birth to adulthood. A specific objective based on this anatomic description is to identify structural changes that relate to the ontogeny of the system used for speech production, mastication, and swallowing. The major issues discussed are (a) the degree to which growth patterns differ for the various parts of this complex system, (b) the degree to which changes in size are accompanied by changes in shape, and (c) how reformation of the overall anatomy relates to the development of speech and other oral motor behaviors. This review emphasizes size and shape changes in the craniofacial skeleton, nasopharynx, lips, tongue, and larynx. Selected information on histology is presented, but the review focuses on gross anatomy. Also included are notes on clinical application and summary tables of selected major sources on anatomic development.

Detailed knowledge of the developmental anatomy of the craniofacial, oral, and laryngeal systems is important for a number of reasons. It has been estimated that more than 25% of Mendelian syndromes have some manifestation in the craniofacial system (Salinas, 1980). Craniofacial dysmorphologies therefore figure prominently in the identification of various genetic syndromes. Reconstructive and ablative surgeries performed on children are planned to take developmental growth patterns into account, and orthodontic management similarly depends on knowledge of growth patterns and adjustments.

In addition, it is often assumed that some aspects of speech development, including speech motor control, are influenced by maturational changes in craniofacial, oral, and laryngeal anatomy. Bosma (1976), for example, proposed that there are two stages of speech production, one based on the vocal tract anatomy of the infant and another based on the quite different anatomy of the older child or adult. Similarly, Thelen (1991) proposed a system dynamic theory of speech development in which anatomic restructuring plays a major role in the emergence of attractor states, that is, stable patterns of motor control. An elaborated system dynamic theory of the development of speech production awaits an understanding of the anatomic development of the vocal tract. Noting that children are functionally different from adults, Stathopoulos and Sapienza (1993) called for models of speech production that are age specific. A major step in formulating such models is the documentation and interpretation of developmental differences in the anatomic systems of speech production. This information also is relevant to the understanding of the various speech and voice disorders that occur in children. For example, Adams (1982), noting that 75% of the reported cases of childhood stuttering develop between the ages of 2 and 7 years, suggests that the "variety of major anatomical and physiologic changes" (p. 171) in this 5-year period have a major role in determining the child's level of fluency.

Sex also is important for physiologically valid models of speech production and for appropriate planning of surgical intervention. Males and females differ in the size and shape of various structures in the aerodigestive tract. Moreover, racial differences have been noted in some structures, such as craniofacial dimensions, but not much is known about racial differences in the vocal tract as a whole.

This review concentrates on anatomic development between birth and young adulthood, but some functional issues are considered in line with anatomic changes. These issues include speech development, clinical assessment, and sex and racial differences in functional properties.

GENERAL ISSUES IN GROWTH AND DEVELOPMENT

Problems in Measuring the Growth of the Craniofacial and Laryngeal Structures

The measurement and description of physical growth are not straightforward. Particularly in the case of the cranium, lips, tongue, vocal tract, and larynx, growth may entail increases in size with concomitant changes in shape. Because increases in size are accompanied by morphologic variations, it can be difficult to identify measurements, especially linear ones, that are stable across age. In the worst case, changes in shape invalidate the use of a given measure across the lifespan.

Among the most abundant data are those on cranial skeletal form, and it is instructive to review briefly the measurement procedures that are commonly used. The traditional method is the Conventional Metrical Approach (CMA), in which measures of linear distances, angles, and ratios are made from either direct or indirect observations. An example of direct observation is the measurement of the living face and cranium with calipers, sometimes called direct cephalometry. An example of indirect observation is roentgenographic cephalography (i.e., x-ray). A major shortcoming with these traditional measures is the lack of stable anatomic reference points. An individual structure often changes in both size and shape and, moreover, may change in its relation to surrounding structures. Lestrel (1989) puts it frankly: "These problems have arisen because CMA was developed for the measurement of regular geometric objects and never was intended for the numerical description of the shape of complex irregular forms" (p. 79).

Metal implants have been used to avoid some of the problems in conventional CMA (Bjork, 1955, 1968; Iseri & Solow, 1995; Moss, Moss-Salentijn, & Ostreicher, 1974), but implants have restricted regions of application. Other alternatives or modifications to traditional CMA include multivariate statistics (factor analysis, principal components, generalized distance, and canonical analysis) and morphometrics (especially Bioorthogonal Grids [BOG], the finite element method [FEM], and Elliptical Fourier Functions [EFF]. Lestrel (1989) discussed the advantages and limitations of many of these methods. In the following review, measurement systems are primarily of these three types: conventional metric, multivariate statistic, and morphometric. Anthropometric points of reference for measures of craniofacial dimensions are illustrated in the Appendix (Figures A1 and A2). In addition, Farkas (1992) provides a brief summary of relevant information on craniofacial measurements. Additional perspectives are described in Allanson (1989); Kapur, Lestrel, Garrett, and Chauncey (1990); Lestrel (1989); Richtsmeier and Cheverud (1986); Smahel and Skvarilova (1988); Thomas, Hintz, and Frias (1989); and Ward (1989).

Framework for Review of Anatomic Development

The information on development of the craniofacial and laryngeal systems is not easily condensed or integrated. In an effort to make the data more easily interpretable, the following outline will be used in summarizing the developmental patterns:

- Overview of vocal tract anatomy in the infant and adult
- Development of craniofacial form (primarily skeletal tissues)
  - Craniofacial complex
  - Mandible
  - Hyoid
  - Dentition
- Soft tissues of the vocal tract
  - Pharynx (overall)
  - Soft palate and nasopharynx
  - Lips and labio-oral region
  - Tongue
- Larynx and related structures
  - Epiglottis
  - Trachea
  - Larynx
    - Thyroid cartilage
    - Cricoid cartilage
    - Arytenoid cartilage
    - Vocal folds and glottis

This organization is followed because it reflects the general pattern of maturation of the anatomy under review. The head completes its growth relatively early and may be regarded as the first of the structures under review to reach its mature form. Enlow (1975) described the cranium as a template on which the face develops. The face matures after the head and is closely followed by the larynx and perhaps the tongue. The general development of each structure will be described in terms of its (a) status in the neonate; (b) growth pattern, including growth spurts; and (c) age of maturity. But some caveats should be noted. In addition to problems of measuring structures that change in shape as well as size, the depiction of growth is complicated by uncertainties in choice of developmental index (e.g., chronologic age, body length, body weight, regional features such as cranial length, or a tissue growth index such as somatic, neural, or lymphatic growth). Unfortunately, the answer is not always clear, and even if it were, the published studies do not uniformly include the necessary subject information to constitute the appropriate groupings. By necessity, this review makes a number of compromises. Data summaries are presented with chronologic age as the developmental index whenever possible, primarily because this was the most uniformly reported subject description. However, it should be noted that changes in body length or height may be the subject characteristic most highly correlated with the growth of most of the structures reviewed in this paper.

Brief notes on recent clinical application or significance are placed at the end of major sections. However, the major purpose of the review is to discuss overall anatomic development, and the clinical endnotes are by no means comprehensive. They are included primarily as suggestions for potential augmentations to assessment procedures. As such, they are best considered as supplements to the systematic assessments of the craniofacial-oral-laryngeal system discussed by Hodge (1991), Love and Webb (1992), Mason and Simon (1977), Morris (1982), and Robbins and Klee (1987).

Developmental Periods

When possible, the following developmental periods will be used in subject descriptions (Valadian & Porter, 1977).

Prenatal Period

This period extending from conception to birth, or about 40 weeks, is subdivided into three stages: an embryonic period from conception to 8 weeks, a middle fetal period from 9 weeks through 24 weeks, and a late fetal period from 25 weeks to birth.

Infancy

Infancy, from the Latin for "incapable of speech," generally refers to the first 2 years of life. This period has three stages: neonatal period (birth to 1 month), infancy proper (1 month to 1 year), and late infancy (the second year of life).

Childhood

Childhood runs from 2 to 10 years for girls and from 2 to 12 years for boys. The preschool years are ages 2 to 6 years, and the school years are ages 6 to 10 years for girls and 6 to 12 years for boys.

Adolescence

The adolescent period is from 10 to 18 years for girls and from 12 to 20 years for boys. The three periods of adolescence are: prepubescence (10 to 12 for girls, 12 to 14 for boys), pubescence (12 to 14 for girls, 14 to 16 for boys), and postpubescence (14 to 18 for girls, 16 to 20 for boys).

OVERVIEW OF VOCAL TRACT ANATOMY IN THE INFANT AND ADULT

Several reviews and atlases of craniofacial and vocal tract anatomy in the infant have been published (Bosma, 1975a, 1975b, 1985; Crelin, 1973, 1976; Fletcher, 1973; Hirano & Sato, 1993; Laitman & Crelin, 1976; Laitman & Reidenberg, 1993; Pierce, Mainen, & Bosma, 1978). The following general description is intended to highlight the major anatomic features of the human newborn, which is the developmental starting point for this review. Many of the features can be seen in Figure 1. For comparison, the vocal tract anatomy of an adult male is shown in Figure 2 and that of a 6-year-old child in Figure 3.

The infant's oral cavity is nearly filled by the broad, flat tongue. The superior boundary of the chamber is formed by a hard palate that is short, wide, gently arched, and frequently covered by five to six roughly formed transverse folds. The infant larynx is positioned high in the neck, approximately at the level of the second or third cervical vertebra. Because of this high laryngeal position, the lingual surface is contained within the oral cavity and does not extend into

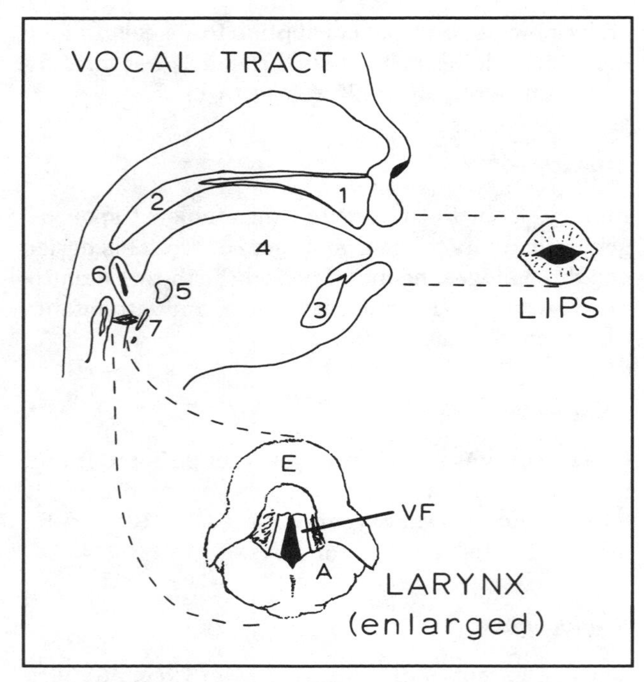

Figure 1. Drawing of a midsagittal section of the vocal tract of a newborn infant. Anterior view of lip configuration also shown. *Key:* 1. hard palate, 2. soft palate, 3. mandible, 4. tongue, 5. hyoid bone, 6. epiglottis, 7. larynx. Enlarged superior view of the larynx: E = epiglottis, A = arytenoid cartilage, VF = vocal folds. Note the following major anatomic features: the tongue fills the oral cavity, the larynx is positioned high in the neck, the epiglottis is close to the soft palate, and the lips have an almost circular shape.

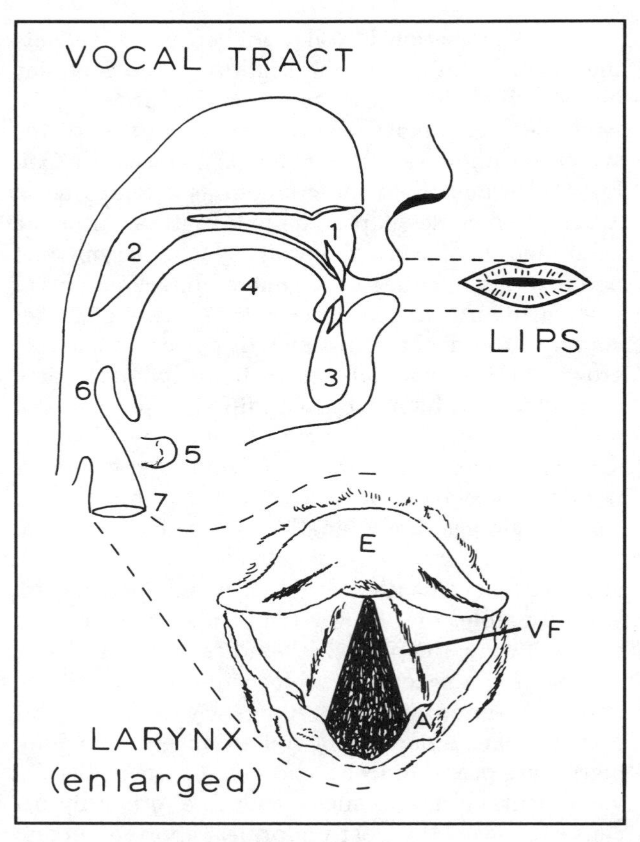

Figure 2. Drawing of a midsagittal section of the vocal tract of an adult. Anterior view of lip configuration also shown. Compare with Figures 1 and 3. *Key:* 1. hard palate, 2. soft palate, 3. mandible, 4. tongue, 5. hyoid bone, 6. epiglottis, 7. larynx.

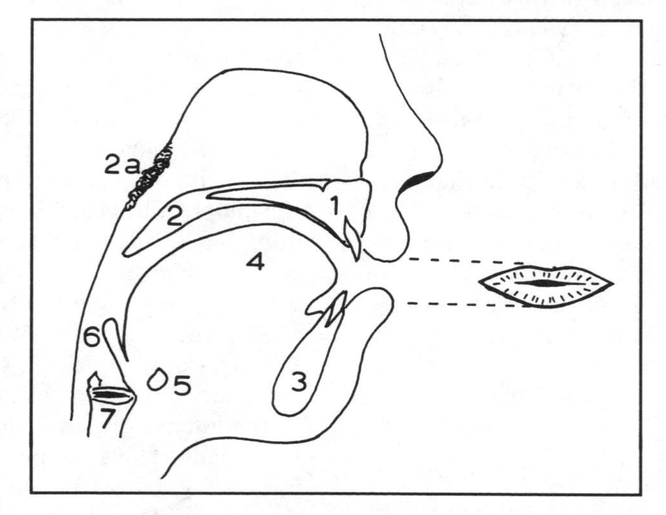

Figure 3. Drawing of a midsagittal section of the vocal tract of a young child (about 6 years of age). Anterior view of lip configuration also shown. Compare with Figures 1 and 2. *Key:* 1. hard palate, 2. soft palate, 2a. pharyngeal tonsil (adenoid), 3. mandible, 4. tongue, 5. hyoid bone, 6. epiglottis, 7. larynx.

the pharynx as it does in the adult. The epiglottis touches, or nearly touches, the soft palate (a configuration referred to as velic-epiglottic engagement). The infant's lips are relatively narrow and rounded. Many of these anatomic features are highly adapted to suckling and no doubt facilitate nourishment in the early phase of life. The tongue has a piston-like action for ingestion of liquids, the lips readily seal around a nipple, the transverse folds of the palate help to grip the nipple, and the velic-epiglottic engagement is thought to enable respiration during suckling as well as to prevent choking. Additional anatomic features are similarly helpful. For example, the neonate's buccal fat pad encased in fibrous connective tissue probably helps to distribute air pressure variations and to keep buccal tissue from being sucked into the space between the gums during suckling and other oral activities. The buccal fat pad also may help to stabilize the infantile mandible during feeding.

An essential feature of orofacial anatomy is that it is one of the most developed muscular systems at birth (Sato & Sato, 1992), providing the infant with the capability for suckling, sucking, and swallowing. Information on these oral behaviors in children has been reviewed by Arvedson and Rogers (1993) and Moore (1988). True suckling is thought to begin at about the 18th to 24th week of gestation. Suckling movements are primarily backward and forward motions of the tongue, the backward phase being more pronounced. Suckling continues until about the 6th postnatal month, when it is replaced by sucking. Sucking differs from suckling largely in that the former has greater activation of the lingual intrinsic muscles to produce vertical tongue movements.

The muscular systems comprise several types, including an agonist-antagonist system related to a hinge joint (mandible), a sphincteric system (lips), and an hydrostatic system (tongue) (Kier & Smith, 1985). These soft tissues are connected largely to bony tissues that follow the classic somatic growth curve (Scammon, 1930). Consequently, the muscles must adapt to growth patterns that extend to at least late adolescence. In fact, it appears that some of the soft tissues of the face, such as the nose and lip, may continue to increase in size into adulthood, at least for males (Nanda, Meng, Kapila, & Goorhuis, 1990). Therefore, although maturity is often assumed to be reached by about 18–20 years, portions of facial growth may continue beyond this period, perhaps even into old age.

The development of the musculature is not unilaterally determined by development of the bony skeleton. In fact, the reverse is often true, as embodied in Moss' (1968) functional matrix theory. As Fletcher (1973) pointed out, the oral cavity is molded by the muscular forces developed internally by the tongue and externally by a muscular ring comprised of the pharyngeal constrictors, buccinator, and orbicularis oris. The oral and perioral muscles are relatively large compared to the bony structures of the hard palate, mandible, and alveolar processes, and therefore can strongly influence the development of the skeletal tissues, which respond to forces placed on them.

Clinical Note:

Magnetic Resonance Imaging of the Vocal Tract

The study of vocal tract structures has been limited by the invisibility of many structures and the lack of suitable imaging procedures. Much progress has been made in the use of magnetic resonance imaging to study vocal tract structures, and it is clear that this method has great potential for both normative and clinical studies. Noteworthy papers include: nasopharynx (Gulisano, Delrio, Montella, Bandliera, & Ruggiero, 1992; Teresi et al., 1987), nasal cavities (Dang, Honda, & Suzuki, 1994; Zinreich et al., 1988), tongue (Lauder & Muhl, 1991; Lufkin, Larsson, & Hanafee, 1983; Vogl et al, 1988), mandible (Christianson, Lufkin, Abemayor, & Hanafee, 1989), masseter muscle (Lam, Hannam, & Christiansen, 1991), larynx (Castelijns et al., 1985, 1987; Sakai, Gamsu, Dillon, Lynch, & Gilbert, 1990; Wortham, Hoover, Lufkin, & Fu, 1986), vocal tract configuration (Baer, Gore, Gracco, & Nye, 1991; Moore, 1992), and correlations between jaw-muscle cross-section and craniofacial morphology (van Spronsen, Weijs, Valk, Prahl-Andersen, & van Ginkel, 1991). A convenient reference atlas in pocketbook form is Lufkin and Hanafee (1989).

DEVELOPMENT OF CRANIOFACIAL FORM

Selected primary sources of information are summarized in Table 1. This is by no means an exhaustive listing of this large literature. Methods used to determine the growth of the cranial base and related bones have been primarily x-ray cephalometry and dissection. The same methods in addition to direct cephalometry have been used to study growth patterns of the face. The color atlas of McMinn, Hutchings, and

TABLE 1. Selected sources for anatomic development of the cranial form (see Table 3 for additional sources on nasolabial features). Studies are listed chronologically. *N* = number of subjects; all ages in years, unless otherwise specified. GA = gestational age.

Source	*N*	Ages (in years)	Comments
Hellman (1932)	1196	5–22	Direct cephalometry with data reported relative to dental development.
Goldstein (1936)	50	2–21	Direct celphalometry at 2-year intervals; also includes data for 50 elderly men.
Brodie (1941)	21	0–8	Serial lateral x-rays (3-month intervals for first year of life; 6-month intervals for 1–5 years, then annual intervals).
Tracy & Savara (1966)	50	3–16	X-ray cephalometry annually for at least 6 years for each subject.
Walker & Kowalski (1972)	802	6–26	X-ray cephalometry in a cross-sectional study of 380 males and 422 females.
Hajnis (1974)	2729	0–18	In German.
Broadbent, Broadbent, & Golden (1974)	32	1–18	Annual standards for facial growth.
Moss et al. (1974)	NR	Varies with subject	Case studies of mandibular growth with logarithmic spiral modeling.
Riolo, Meyers, McNamara, & Hunter (1974)	—	—	Atlas of craniofacial growth
Snyder et al. (1977)	4127	2–18	No longer in print.
Blum & Weber (1979)	—	—	Angular invariants explored with the Broadbent-Bolton cephalometric standards.
Bishara, Peterson, & Bishara (1984)	35	4.5–17	X-ray cephalometry bienially between 4.5 and 12 years and annually through age 17. Additional records at adulthood (mean age of 25.5 years).
Lavelle (1984)	90	12–15	X-ray cephalometry of female subjects representing the three major Angle molar occlusal categories.
Merlob, Sivan, & Reisner (1984)	198	27–41 weeks GA	Physical measurements from 87 term and 111 preterm infants. Of the 27 measures reported, 14 pertain to craniofacial features.
Richtsmeier & Cheverud (1986)	179	4–15	19–20 males at each of the following ages: 4, 5, 7, 8, 9, 10, 12, 13, and 15 yrs (all males). Finite element analysis of craniofacial complex.
Mehes (1987)	366	Newborn	Anthropometric standards for head length and width.
Nickel et al. (1988)	49	0–20	Examination of temporomandibular joint from osteologic remains.
Smahel & Skvarilova		Adult	X-ray cephalometric study of cranial interrelations.
Lestrel (1989)	—	—	Discusses approaches to mathematical modeling.
Omotade (1990)	508	Newborns	Direct cephalometry with transparent grid positioned over subject's face. Data on 252 white newborns born in Cardiff, Wales and 256 black newborns born in Ibadan, Nigeria.

(continued)

Source	*N*	Ages (in years)	Comments
Farkas et al. (1992a)	1537	1–18	Measures of head size, including height, length, width, and circumference.
Farkas et al. (1992b)	1594	1–18	Measures of face height and width, mandible height, width, and depth.

Logan (1981) is particularly helpful in understanding the relations among the cranial and facial bones, as well as their relations to soft tissues.

Craniofacial Complex

The major skeletal elements of the cranium and face are illustrated in Figure 4 in both frontal and lateral views. The adult skull consists of the craniofacial complex and the mandible. The former is composed of the cranium, which surrounds the cranial vault, and the face, which includes the forehead as well as the skeletal framework of the eyes, nose, and mouth (Dickson & Maue-Dickson, 1982). The cranium and face are closely located anatomically but dissimilar in growth patterns. The cranial and facial anatomy of neonate and adult are shown in Figure 5. The cranium is proportionately larger at birth than the facial or body skeleton. Scammon (1930) considered the cranium to follow the neural growth curve, but regarded the middle and lower anterior regions of the face to follow the general somatic growth curve. Between the ages of 10 and 20 years, the cranium grew only about 4%, whereas some facial regions grew about 35%. The cranium often is viewed as essentially mature in size by the age of 6 years, but the face at that age is relatively infantile in both size and shape. Melsen and Melsen (1982) described the brain as having nearly adult size at 5–6 years. The divergent growth patterns between cranium and face require that facial development be essentially liberated from cranial development. The anatomic junction between cranium and face is the basicranium, composed of a midline cranial base and three cranial fossae (for prenatal development of the basicranium, see Kjaer, 1990a, 1990b). The fossae grow in relation to the brain, but the midline cranial base is much more independent of brain growth.

This review is restricted to development from birth to young adulthood. Embryology, although clearly relevant, is outside the scope of this paper. A good source for readers who seek a general review of craniofacial embryology is Sperber and Tobias (1989).

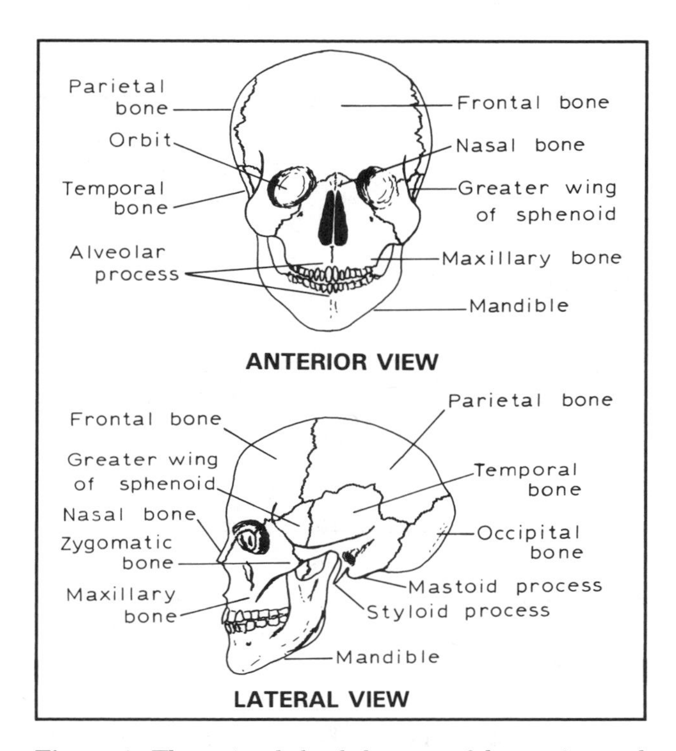

Figure 4. The major skeletal elements of the cranium and face, seen in frontal and lateral views.

The midline cranial base reaches from the foramen magnum to the foramen caecum. Its bony components are the basioccipital, body of the sphenoid, and cranial ethmoid (mesethmoid). Figure 6 is a basal view showing how the sphenoid, occipital, and temporal bones form the base of the skull; the ethmoid bone is not seen in this perspective. Some authors extend the midline cranial base to the nasion, in which case it includes the frontal bone as its anterior margin. Except for the frontal bone, the constituents take their origin in the cartilage of the chondrocranium. The occipital and sphenoid components are separated by the spheno-occipital synchondrosis, an important cartilaginous growth center of the craniofacial complex. Copray, Jansen, and Duterloo (1986)

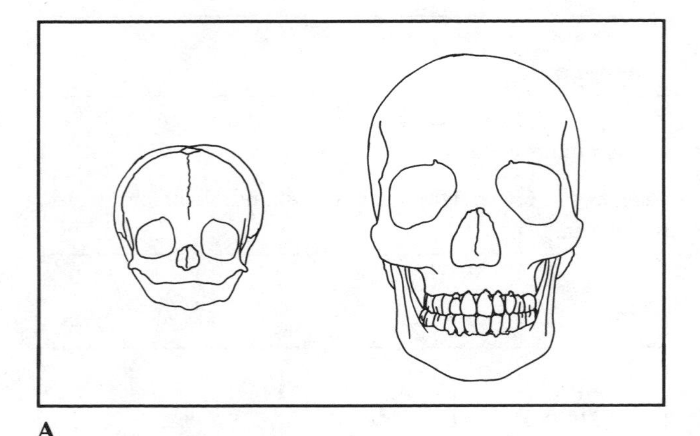

A

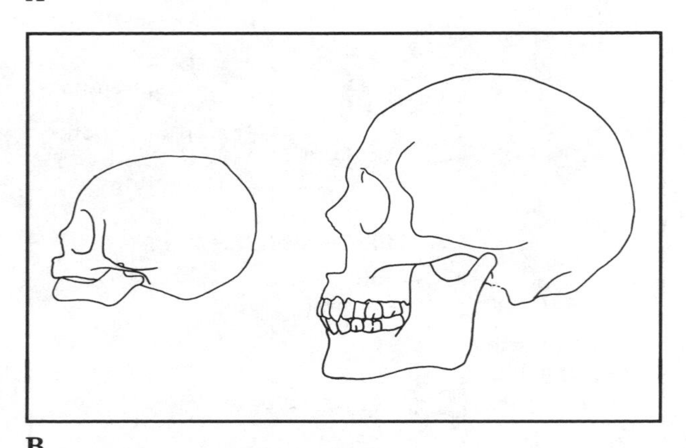

B

Figure 5. A. Frontal view of cranial and facial bones of infant (*left*) and adult (*right*). **B.** Lateral view of cranial and facial bones of infant (*left*) and adult (*right*).

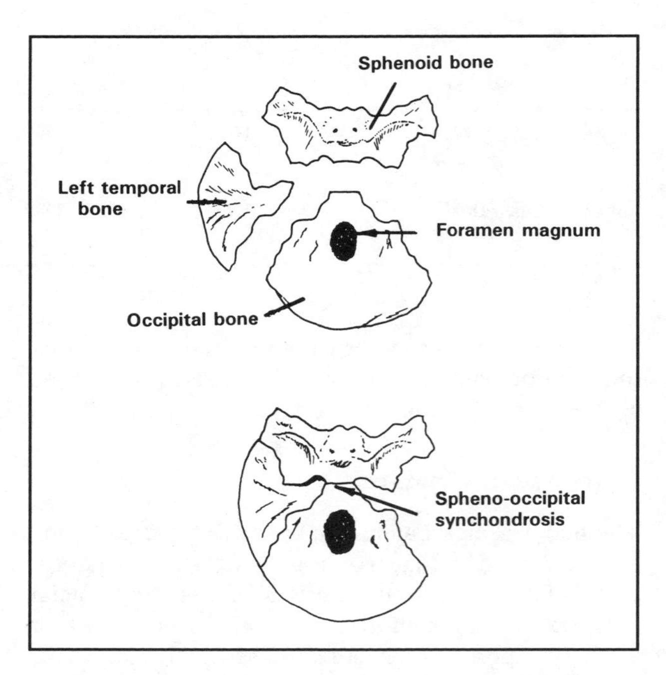

Figure 6. Basal view of the base of the cranium, showing (*top*) the disarticulated sphenoid, temporal, and occipital bones; and (*bottom*) the union of the bones including the spheno-occipital synchondrosis.

emphasized that the synchondrosis is a bioelastic, pressure-bearing joint related to the main load-carrying region of the head. The synchondrosis persists until 12–18 years of age, closing later in boys than in girls (Crelin, 1973; McKearn & Stewart, 1957; Powell & Brodie, 1964). The midline cranial base is susceptible to alterations in Down syndrome, achondroplasia, and a variety of other conditions. Laitman and Crelin (1976) noted that the basicranial region in human infants is similar to that in monkeys and apes. The unique human anatomy of the craniofacial system and vocal tract is in large part a result of an increasing curvature of the basicranium with descent of the tongue to form the superior-anterior pharyngeal space. Laitman and Crelin estimated that these changes are essentially complete by the age of 4 years.

At birth, the normal palatal morphology is round, but sagittal growth, especially in the posterior region, causes a change to oval morphology by about 3 months of age (Kramer, Hoeksma, & Prahl-Andersen, 1992). In addition to this lengthening of the anteroposterior palatal axis, the measurements of Kramer et al. show that the arch depth of the palate increases about 4.2 mm over the first 3 months of life.

The maxillae and the palatine bones constitute the maxillary complex. The growth centers of this complex are the frontomaxillary, zygomaticomaxillary, and zygomaticotemporal sutures (Figure 4). The sutures are arranged so that the pattern of growth of the maxillary complex is downward and forward relative to the cranium.

Melsen and Melsen (1982) provided a chronology of development of the palatine bones and the following summary is based largely on their account. The palatine bones in the infant are separated from both the maxilla and the pterygoid process of the sphenoid. Consequently, the palatine bones are capable of movement relative to the adjacent bony structures. The contacts become more intimate in the juvenile but the bones can be disarticulated with some difficulty. By adolescence, they cannot be separated without fracture. The growth pattern of the maxillae and palatine bones is as follows. The anterior part of the maxillae descends first, followed by the posterior section, which consists of the horizontal process of the palatine bones articulating with the maxillae and pterygoid processes. The descent is accomplished by growth at the sutural line and remodeling, with resorption of bone on the nasal surface and apposition on the oral side. The palatine bones follow the descent of the anterior maxilla by means of increased remodeling.

The facial skeleton develops to transfer the forces produced during mastication to the base of the skull. In this sense, the skeleton can be likened to a buttress that transfers forces around the various cavities of the skull. These forces may be highly influential in the anatomic development of the bony framework, determining regions of resorption and apposition of bone.

Current theory emphasizes the relations between soft and hard tissues in the development of the craniofacial system. One of the most influential theories in this respect was Moss' (1968) *functional matrix theory,* which Ranly (1988) succinctly expressed as follows:

> There is no direct genetic influence on the size, shape, or position of skeletal tissues, only the initiation of ossification. All genetic skeletogenic activity is primarily dependent upon the embyronic functional matrices. (p. 147)

Moss proposed that the individual functions of the human head can be related to various functional cranial components. The component associated with the function of speech consists of the lips, teeth, tongue, soft palate, oral and nasal cavities, and other oral structures involved in any aspect of speech production. Each component has two major parts, a *functional matrix,* consisting of the soft tissues and spaces that perform a function, and a related *skeletal unit* that supports and protects its functional matrix through biomechanical actions. The skeletal unit for speech articulation includes principally the maxilla and mandible. Of course, the same structures are involved in other functions including mastication and deglutition. The various functions would determine the eventual size and shape of the skeletal components.

Cephalometric data on the face obtained from both direct and indirect methods are abundant and will not be reviewed in detail here. Table 1 is a selected representation of the numerous studies in this area. The papers of Farkas and colleagues (Farkas, Katic, Hreczko, Deutsch, & Munro, 1984; Farkas & Posnick, 1992; Farkas, Posnick, & Hreczko, 1992a, 1992b) are noteworthy because they provide contemporary data for large numbers of North American subjects for a variety of facial dimensions. Some of the data from Farkas and colleagues are summarized graphically in Figures 7 (dimensions of the head), 8 (measurements of the face), and 9 (measurements of the nasolabial region).

The work of Farkas and colleagues is an example of CMA, which, as explained earlier, is the derivation of distances, angles, and ratios. However, CMA has been questioned for its suitability to structures that change in shape as well as size. A variety of statistical and morphometric analyses have been undertaken as alternative systems for measurement and analysis. Richtsmeier and Cheverud (1986) studied the development of craniofacial regions in males aged 4 to 18 years, using a finite element analysis. Some major results were as follows: (a) the upper face has a steady increase in size and a steady change in shape; (b) the middle face changes less in shape than the other elements in the analysis; (c) the lower face has a period of rapid growth between 7 and 9 years; (d) the palate changes more in shape than size until age 12, after which shape changes become less influential; and (e) the basisphenoid had average rates of change in size and shape.

Summary

The facial skeleton is contiguous with the cranial skeleton but follows a very different maturational schedule. Whereas the cranium approaches adult size relatively early in childhood, the facial bones continue to grow until adolescence and possibly even adulthood. These divergent growth patterns result in a complex developmental picture which can be understood through an appreciation of the growth centers of the cranial and facial bones. The basicranium is particularly important in understanding facial growth and the development of the human facial profile and vocal tract. The palatal and maxillary bones grow forward and downward relative to the cranium. Substantial sets of data have been obtained on the growth and development of the facial profile. These data on facial growth hold particular relevance to the specialties of dentistry and reconstructive surgery but they also contribute to the general picture of craniofacial growth of interest to several other specialties, including speech-language pathology.

Mandible

One of the remarkable differences between adult and infant skulls is the proportionately small mandible of the infant (as pictured in Figures 1, 5, and 10). The mandible grows considerably in childhood and its de-

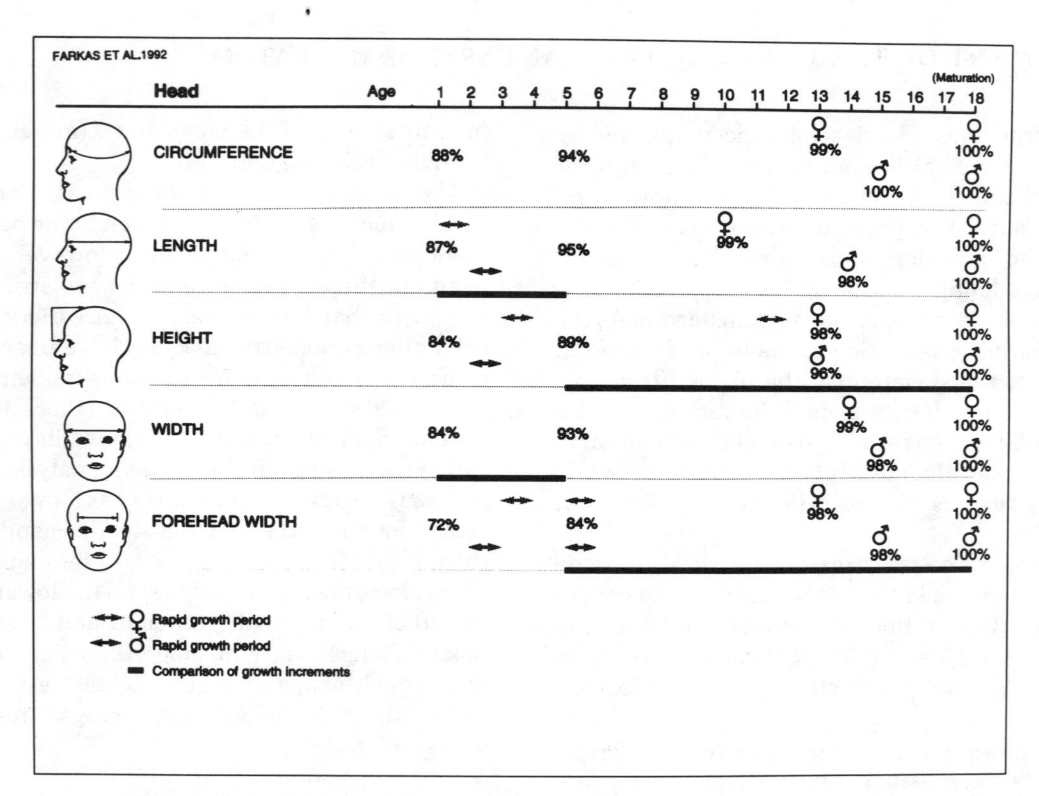

Figure 7. Timetable for development of the head, based on direct cephalometric data. Drawn with permission from measurements in Farkas, Posnick, and Hreczko (1992a).

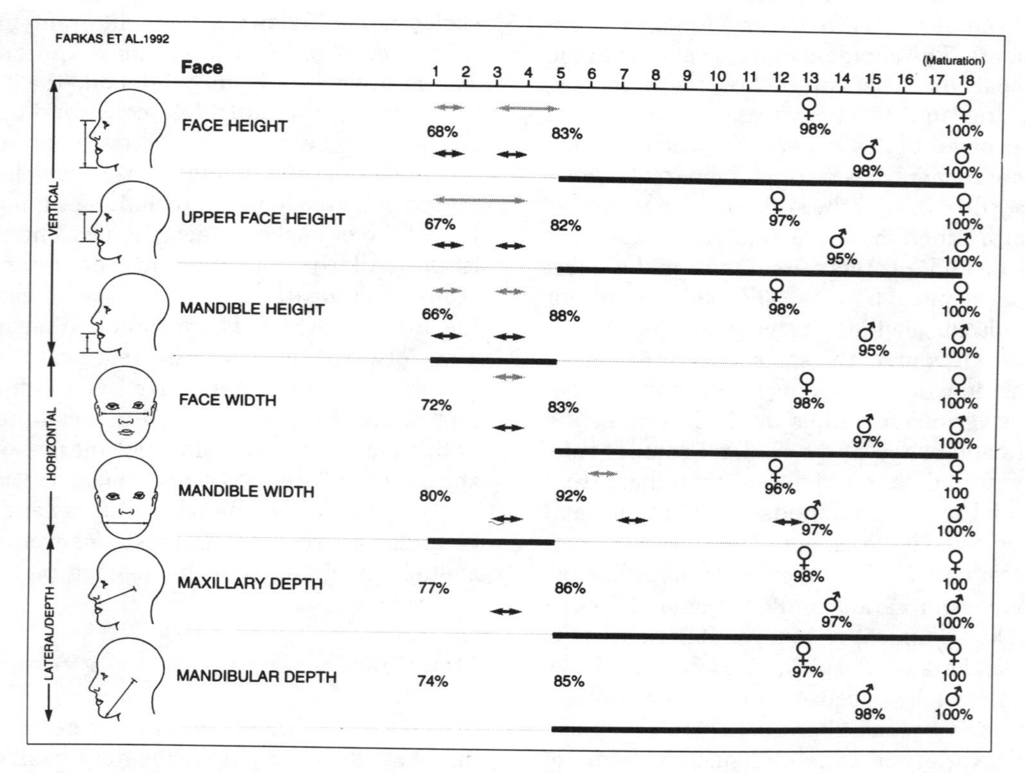

Figure 8. Timetable for development of the face, based on direct cephalometric data. Drawn with permission from measurements in Farkas, Posnick, and Hreczko (1992b).

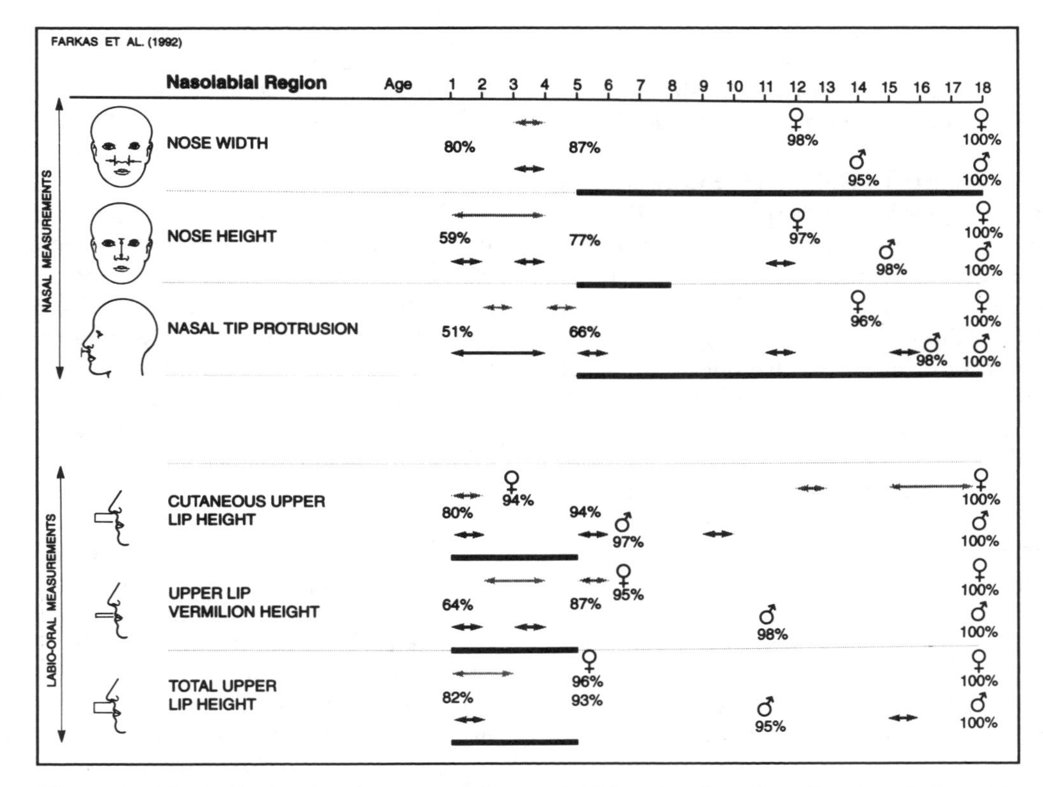

Figure 9. Timetable for development of the nasolabial region, based on direct cephalometric data. Drawn with permission from measurements in Farkas, Posnick, Hreczko, and Pron (1992).

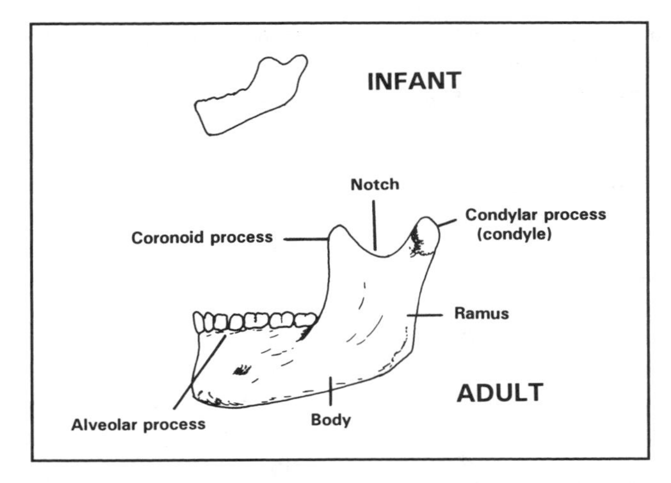

Figure 10. Drawings of the mandible in infant and adult.

velopment contributes importantly to the maturation of the face. The following summary is based largely on Scott's (1976) description of the general processes of mandibular development.

One aspect of mandibular growth is the conversion of growing cartilage to bone. The two primary sites for this conversion are the midline symphysis (active until about the first year) and the condyles (active until adulthood). The other aspect of mandibular growth is deposition and resorption of bone. Postnatal growth through bone deposition is accomplished principally at the alveolar margin, posterior border of the ramus, and upper edge of the ramus (including the condyle, coronoid process, and mandibular notch). Primary sites of resorption are the anterior border of the ramus and the mesial surface of the body and ramus.

Harris (1962) reported a childhood growth spurt in mandibular development, but Meredith (1961) characterized the growth pattern as having a smoothly declining rate of increment. Tracy and Savara (1966) interpreted their data as supporting Meredith's conclusion, but they did observe a circumpuberal maximum in growth rate beginning at about 9 years of age. One of the largest studies on mandibular growth is Walker and Kowalski's (1972) cross-sectional investigation of 800 normal white American subjects. Whereas the mandible grew slowly and steadily in females over the age range of 8–20 years, it grew rapidly in males between 12 and 16 years. Woodside's (1969) findings on mandibular growth, as reported in Brulin and Talmant (1976), reflect growth spurts at approximately ages 6–7, 8–10, and 14–16 years for males, and 6–8 and 11–13 years for females. Additional data on sexual dimorphism in mandibular growth were derived from the serial lateral cephalograms of the classic Bolton-Brush Study (Ursi, Trotman, McNamara, & Behrents, 1993). Ursi et al. concluded that the effective lengths of both the maxilla and mandible were similar in males and females until 14 years of age. After that age, the measurements for females changed little but increases were noted for males. Interestingly, Nayeem (1992) described the mandible as "the only bone by which both sex and age can be detected," apparently basing this conclusion on studies such as that of Wowern and Stoltze (1979) in which age and sex differences were observed in the mandible but not the second metacarpal bone.

Development of the mandible may be accomplished largely by growth in size and little or no change in shape (Figure 10). Moss, Moss-Salentijn, and Ostreicher (1974) advanced three interrelated conclusions about mandibular growth: (a) it is allometric (i.e., it fits the relative growth curve, $y = bx^e$); (b) it can be described by a logarithmic growth curve; and (c) it is gnomonic, meaning that the mandible changes only in size, not in shape. Consistency in shape was also demonstrated by Webber and Blum (1979), who remarked on the consistency of the mandibular outline form across different age groups. Ontogenic gnomonicity has a parallel in studies of mandibular variation across individuals. Lavelle (1984) reported that, for females between 12–15 years of age, mandibular shape is less variable than size, for subjects both within and across Angle occlusion classes. It seems possible, then, to specify a general anatomic shape of the mandible that is preserved across age and occlusal stages. (See also Lavelle, 1985, and Lavelle and Greenwood, 1985, for information on mandibular shape.)

The mandible is the only movable part of the adult craniofacial skeleton, and its kinematic characteristics are important clinically. The movement range of the mandible in the adult allows for a maximum opening of about 55–60 mm (33° of angular variation) in men and about 50–55 mm (35° of angular variation) in women (Pullinger, Liu, Low, & Tay, 1987). In its normal resting position, the jaw maintains a small and consistent incisal opening, or "freeway space," of about 3 mm in adults (Niswonger, 1934). These functional properties are useful in evaluating the general status of mandibular motor regulation, and they define the envelope of jaw movement.

Some aspects of mandibular growth are closely related to development of the dentition. Nickel, McLachlan, and Smith (1988) commented that the eminence of the temporomandibular joint reaches more than half of its mature size and has a mature morphology by the time the primary dentition is completely erupted. They also observed that the maximum velocity of development of the eminence occurred before the age of 3 years. Termination of growth was estimated to occur by the mid to late teens.

Summary

Although studies of mandibular growth have been interpreted to support either a smooth growth process or one marked by circumpuberal growth spurts, the bulk of the evidence seems to support the latter conclusion, especially for males. In general, the mandible follows the skeletal growth pattern and reaches its adult size by about 18 years of age in males and somewhat earlier in females. An important simplifying assumption in characterizing the growth of the mandible is that changes in size are accomplished with relatively minor changes in shape across age or occlusal stages.

Clinical Note:

Masticatory System

Morris and Klein (1987) remarked that the ability to stabilize the mandible is a prerequisite for the development of skilled and refined tongue and lip movements. As mentioned earlier, the infant's sucking pads contribute to such stabilization. However, premature infants are at a disadvantage because they often lack these pads. Manual stabilization by the caregiver serves the function of enhancing nutritive intake during suckling, as well as providing the infant with stability for the development of oral motor skills, such as dissociation of tongue from jaw movements.

The relation of the gum pads and erupting teeth stabilizes during the first year of life, and the first primary molars achieve occlusal contact by the age of 16 months (Widmer, 1992). Following establishment of this occlusal contact, the jaw typically closes to the same position. Children who exhibit open-mouth posture have a slower pattern of maxillary growth than children with a closed-mouth posture (Gross, Kellum, Franz et al., 1994; Gross, Kellum, Michas et al., 1994). Therefore, mandibular function influences the development of the maxilla.

Occlusion, asymmetry of development, and function were emphasized in a series of studies that examined the relationship between anatomy and speech production. Laine, Jaroma, and Linnasalo (1987) reported that "deviant traits in the anterior part of the dentition" (p. 84) are linked to difficulties in producing medioalveolar consonants such as the /s/. Pahkala, Laine, and Lammi (1991) concluded that symmetrical eruption of the permanent teeth and individual timing pattern are more important than the mean values of tooth eruption. They also reported reduced risk of specific speech sound production errors in 7-year-old children with advanced eruption of some permanent teeth. Pahkala (1994) reported that the prevalence of craniomandibular disorders increased with growth, although individual signs and symptoms fluctuated. There is evidence that jaw opening may be related to articulatory disorders, especially in boys (Ettalo-Ylitalo & Laine, 1991). The dynamic technique of computerized axiography was used by Bigenzahn, Piehslinger, and Slavicek (1991) to identify orofacial dysfunctions including orofacial imbalance. Hale et al. (1992) reported that single-syllable diadochokinetic rates were slower in children with an open-mouth resting posture, possibly indicating a relation between postural and kinematic behaviors.

The sensorimotor development of the mandibular system may extend over the human lifespan. Smith, Weber, Newton, and Denny (1991) concluded from studies of jaw-closing reflexes that there is a continuous evolution of oral sensorimotor systems throughout the lifespan. Additionally, they reported that oral-motor reflexes develop in concert with complex motor skills. Therefore, reflexes are not simply invariant response patterns from which complex motor acts can be constructed. This view could have important clinical implications for a lifespan perspective on sensimotor functions of the oral system. The data of Smith et al. (1991) and Wood and Smith (1991) indicate that the age of 7 to 8 years may be a transitional period for cutaneous oral-motor reflex development. Before this transitional period, the oral-motor system is not as responsive to mechanical stimuli.

Hyoid

The horseshoe-shaped hyoid bone is the supportive structure for the base of the tongue as well as as the upper larynx. The hyoid and mandible are the skeletal pedestals for the tongue and probably contribute to the piston-like actions of the tongue in infantile suckling and sucking (Fletcher, 1973). Crelin (1973) described the newborn hyoid to be 5 mm high, 2 mm thick, and to have a distance between the greater cornua (horns) of 3 cm. The hyoid bone in the average normal subject is aligned roughly parallel to the lower border of the mandible. During development, the hyoid descends in relation to the mandible and styloid process (Bosma, 1975b). In the infant, the body of the hyoid is situated at the junction of the second and third cervical vertebral bodies, whereas in the adult it is situated at approximately the level of the body of the fourth cervical vertebra (Bench, 1963; Ardran & Kemp, 1972). Gallati (1899; cited by Fink, Martin, & LaVigne, 1975) stated that in the first year of life the hyoid overlaps the highest point of the thyroid cartilage anteriorly. This overlap, according to Gedgoud (1900, cited by Fink et al., 1975), is 1–3 mm up to 3 months of age. A cephalometric study by Bloquel, Laude, Lafforgue, and Devillers (1990) on children aged 6.5 to 17.5 years, revealed the descent of the hyoid bone (measured from the superior-anterior tip of the hyoid bone to the tangent line of the palatine laminae) to be up to 33 mm, with some anteroposterior variations across subjects. (The authors reported such variations to influence the posterior modifications of the palate.) Bench's (1963) measurements from lateral cephalometric films on 165 subjects, ages 2 years to adult, revealed sex differences in the vertical descent of the hyoid especially between the ages of 12 to 18 years. The descent of the hyoid is reportedly completed by age 19 and thus follows the general skeletal growth pattern. A stable anteroposterior relation between the hyoid and cervical vertebrae is seen until puberty in the form of a fixed hyoid-vertebral distance (Brodie, 1971; King, 1952).

Summary

The hyoid, a uniquely floating bone (unattached to other bones) in the human anatomy, provides skeletal support for the tongue and skeletal suspension for the larynx. A remarkable feature of the development of the hyoid is its descent from a superior position in the infant (approximately the second or third cervical vertebra) to a more inferior position in the adult (fourth cervical vertebra). The adult position is reached by the age of about 16 years.

Clinical Note:

Hyoid Bone

Tourne (1991) discussed the importance of the hyoid bone in relation to oropharyngeal growth and physiological adaptations to decreased patency of the pharyngeal airway. Ardran and Kemp (1972) emphasized the importance of age of subject in assessing the relation between the mouth cavity and the position of the hyoid bone. The hyoid bone is important as a site of muscle attachments, with over 20 muscles either inserting on, or originating from, this bone (Zemlin, 1988). Because the body of the hyoid bone typically moves superiorly and anteriorly during swallowing, motion of the hyoid bone can be used in the assessment of swallowing.

Dentition

The most striking developmental feature of the teeth is the emergence of the deciduous dentition in the infant and young child and the eventual replacement of the deciduous dentition by the permanent dentition (Figure 11). Scott (1976) pointed out that both the deciduous and permanent dentition are composed of the various dental units implanted in the alveolar processes of the maxilla and mandible. Each tooth consists of the crown, neck, and root. The crown is the exposed portion of the tooth. It contains a cavity filled with dental pulp and is covered with enamel. The neck is the constricted region between the crown and the root. The root is embedded in the alveolus (tooth sockets) and is held by fibers of the periodontal ligament. Scott emphasized that, by virtue of these transeptal fibers, the teeth have a serial chain-like arrangement and therefore have a collective unity that must be appreciated to understand dental growth and processes involved in orthodontia. For example, one reason for orthodontic treatment is crowding of teeth. Gaps between teeth also occur. These spaces are called diastemata or primate spaces, because many animals have marked separations in their dental arches (e.g., spacing between the upper incisors and canines and between the lower canines and first molars). Widmer (1992) pointed out that diastemata are commonly present in early primary dentition, and their absence indicates that the permanent teeth may be crowded.

The formation of both the primary and secondary teeth begins in utero. The three important stages in dental development are the initial mineralization or calcification, crown completion, and root completion. Calcification of the deciduous teeth begins in utero, whereas calcification of permanent teeth begins postnatally (Widmer, 1992). At birth the indented gum pads indicate the position of the developing teeth. As shown in Figure 11, the primary teeth begin to emerge at about 6 months.

Liversidge, Dean, and Molleson (1993) measured tooth length in 63 individuals between birth and 5.4 years, from remains of an archeological population, and plotted tooth length against individual age. They reported differences in growth rate between deciduous and permanent teeth with growth rate of deciduous teeth being more rapid. They also observed differences in rate of growth and dentition type. The growth of the incisors and canines was faster than the growth of molars. However, their findings showed that, for both deciduous and permanent teeth, crown completion preceded root completion for all teeth types. Liversidge et al. concluded that, during early development, direct measurement of crown height may be a fairly accurate means of estimating age. However, Demirjian (1980) maintained that only radiographic studies, which measure dental calcification, provide an accurate measure of maturity, and hence it is more appropriate to use dental maturity scores than dental age or chronological age.

Wide individual variations are seen in the eruption of both the deciduous and permanent teeth. In addition, Demirjian (1980) pointed out that, although there are known differences in the development of mandibular and maxillary teeth, those developmental differences are not clearly understood, and the left mandibular dentition is typically used to represent overall dental development. The normal variation in eruption can be 6 to 12 months on either side of the mean age for the deciduous teeth and 12 to 18 months on either side of the mean age for the permanent teeth (Bernat, 1992). Total eruption time is about 20 months for the deciduous dentition and about 14 years for the permanent dentition.

In addition to the more variable age of eruption of the permanent teeth compared to the deciduous

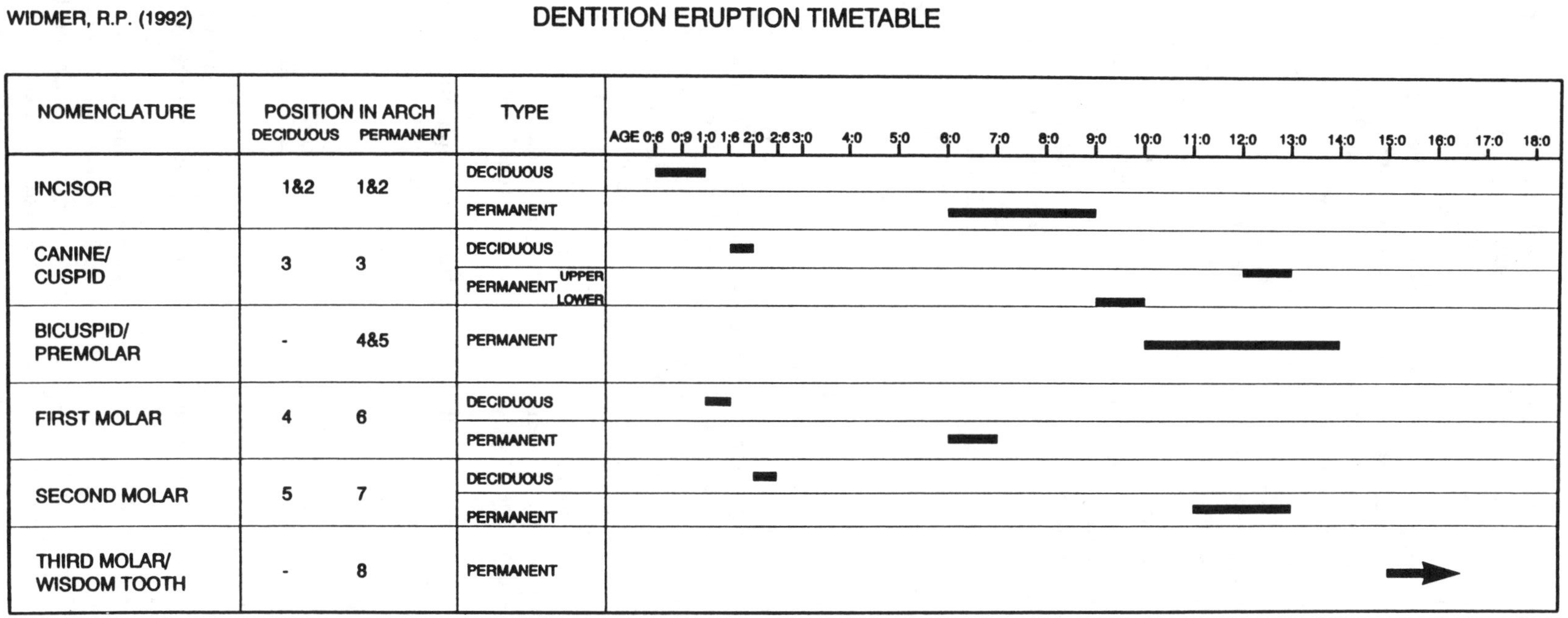

WIDMER, R.P. (1992)

DENTITION ERUPTION TIMETABLE

NOMENCLATURE	POSITION IN ARCH DECIDUOUS	POSITION IN ARCH PERMANENT	TYPE		AGE 0:6 0:9 1:0 1:6 2:0 2:6 3:0 4:0 5:0 6:0 7:0 8:0 9:0 10:0 11:0 12:0 13:0 14:0 15:0 16:0 17:0 18:0
INCISOR	1&2	1&2	DECIDUOUS		
			PERMANENT		
CANINE/ CUSPID	3	3	DECIDUOUS		
			PERMANENT	UPPER	
				LOWER	
BICUSPID/ PREMOLAR	-	4&5	PERMANENT		
FIRST MOLAR	4	6	DECIDUOUS		
			PERMANENT		
SECOND MOLAR	5	7	DECIDUOUS		
			PERMANENT		
THIRD MOLAR/ WISDOM TOOTH	-	8	PERMANENT		

Figure 11. Chronology of the deciduous and permanent dentition. Based on data in Widmer (1992). Used with permission.

teeth, gender and racial differences also have been reported (Demirjian, 1980; Diaz, Maccioni, Zedda, Cabitza, & Contis, 1993; Widmer, 1992). Demirjian (1980) reported that girls are 1 to 6 months ahead of boys in both calcification and emergence of the permanent teeth. However, calcification of the deciduous teeth does not show sexual dimorphism, and the literature on their emergence is more controversial. Kerr, Kelly, and Geddes (1991), who studied 57 subjects aged 9.3 to 19 years, observed sexual dimorphism in teeth surface area, with measurements being smaller for females. With respect to racial differences, Demirjian (1980) reported the emergence of the deciduous teeth to be delayed in Asiatic and African children as compared to Caucasians. However, the reverse is true for permanent dentition, where the teeth of Caucasians erupt later than those of most other races. Population differences also have been reported in dental maturation (calcification, not emergence). For example, Diaz et al. (1993), who studied dental development in 382 children from Sardinia, reported Sardinian children to be delayed in dental maturation as compared to North American children.

Frequently cited tables of the normal chronology of dental development are those reported by Behrman and Vaughan (1987) and Gorlin, Pindborg, and Cohen (1976). The emergence of the primary dentition and its eventual replacement by the secondary dentition are important events in oral morphology and function. The alveolar bone develops once all the primary teeth have erupted. The alveolar bone development increases both facial height and palatal height (Widmer, 1992). Because the teeth permit forceful masticatory action, they induce the child to generate oral forces that may influence the general development of the orofacial skeleton.

Tongue volume is another morphologic factor that maintains dental arch size, and thus determines in part its boundary (Tamari, Shimizu, Ichinose, Nakata, & Takahama, 1991). A study of 74 Japanese adults with normal occlusion by Tamari et al. showed that lower dental arch sizes were correlated significantly with tongue volume, with stronger correlations at the more posterior parts of the dental arch where tongue mobility is more limited. Although to our knowledge this relationship has not been studied in infants and children, it is likely to be present in them as well. As the teeth erupt, they in turn specify boundaries for tongue position. For instance, the eruption of the incisors in the infant results in a retraction of the tongue, discouraging the lingua-labial contact often seen in edentulous infants. As a consequence, the infant is more likely to produce lingua-alveolar than lingua-labial consonants (Kent & Miolo, 1994). Data on the growth of the maxillary and mandibular dental arches were obtained by Foster, Grundy, and Lavelle (1977). In the regions of the arches anterior to the first permanent molars, growth peaks were observed between 2 and 3 years and 7 and 8 years in the maxilla and between 2 and 3 years and 5 and 6 years in the mandible. The more posterior regions of the arches had growth peaks between 6 and 8 years in the maxilla and between 9 and 10 years in the mandible.

Dental maturation was previously believed to correlate with skeletal maturation, but a number of studies have shown this correlation to be very low (Demirjian, 1980). Data compiled by Lewis (1991) showed only moderate associations between dental ages and left hand-wrist skeletal ages. In some children, the dental and skeletal ages differed by as much as 36 months. Widmer (1992) reviewed studies of genetic influences on dental eruption pattern and teeth size and reported that the connections found between dental eruption time and skeletal maturity, body height, or psychomotor maturity were small.

Summary

Although many processes are involved in the development of the human dentition, tables of the normal chronology of teeth eruption are perhaps of greatest general relevance in clinical application and the general understanding of the development of the craniofacial complex. However, large individual variations are seen in dental development, and care should be taken in drawing inferences as to developmental or clinical status. The large individual variation also complicates the identification of differences related to gender or race. Eruption of the teeth holds many important implications for further remodeling of the craniofacial complex and for the development of speech.

Clinical Note:

Dentition

Enamel defects may be an important biological marker for deafness and other neurologic disorders. Murray, Johnsen, and Weissman (1987) described an association among hearing loss, neurologic impairment, and enamel defects of the primary teeth. Presumably, the association between hearing loss and tooth enamel reflects similarities in the embryologic derivation and development of otic and dental structures. Widmer's (1992) paper is a brief summary of the normal development of the dentition and common disorders. He commented that environmental influences, such as oral habits of finger sucking, can have marked effects on the relation between the maxillary and mandibular teeth.

DEVELOPMENT OF SOFT TISSUES OF THE VOCAL TRACT

Overall Structure of the Pharynx

The pharynx typically is divided into an upper region (nasopharynx or velopharynx), a middle region (oropharynx), and a lower region (laryngopharynx). Information on pharyngeal dimensions in the infant is available in Crelin (1973). The newborn pharynx was estimated to be approximately 4 cm long. The nasal part was considered to be narrow and about 20 mm long, curving gradually to join the oral part without any sharp line of demarcation. At about 5 years of age, the posterior walls of the nasal and oral parts meet at an oblique angle. At puberty, they join almost at a right angle. The length of the adult pharynx is approximately 12 cm long, or three times as long as the infant's pharynx. Pharyngeal growth is almost entirely in the vertical dimension, with little change in the anteroposterior dimension (Bergland, 1963; Handelman & Osborne, 1976; King, 1952; Tourne, 1991). The vertical growth is determined by the extent and direction of growth at the spheno-occipital synchrondrosis and the cervical vertebrae. (Tourne, 1991, gives an informative summary.) The upper pharynx is developmentally more complex than the lower part and has a longer period of morphologic change extending into postnatal life (Fletcher, 1973; Kingsbury, 1915).

Elongation of the pharynx is particularly notable in the adult male, and this anatomic feature distinguishes men from women and men from young male children. Pharyngeal length is an important factor in accounting for gender and age differences in formant frequencies (Kent, 1976).

Soft Palate and Nasopharynx (Velopharynx)

Selected primary sources of information on the tissues of the nasopharyngeal region are summarized in Table 2. Kahane and Folkins (1984) provide helpful photographs of dissections of the nasopharyngeal anatomy.

A frequently mentioned feature of the vocal tract anatomy of the human infant is the engagement of the velopharynx and larynx, manifested by a proximity of the uvula and epiglottis (Crelin, 1973). This arrangement is quite common among mammals and apparently enables breathing to continue even during feeding. The anatomic separation of palate and epiglottis in humans carries an attendant risk of choking but may support the rich phonetic repertoire of speech (Lieberman, 1984). Fletcher (1973) credited the upper pharynx with an early specialization for phonation, noting that, during phonatory activities, pharyngeal sphincteric action occurs in the nasopharynx but ceases in the lower pharynx. He also commented on developmental changes in the spatial relation of the palate to its extrinsic musculature. Because the developing palate descends relative to the origin of the levator veli palatini muscle on the basicranium, this muscle serves as a palatal tensor in the infant but later becomes a palatal elevator. Similarly, because the hard palate descends relatively more during development than the hamular process of the pterygoid palates, the tensor veli palatini muscle depresses the palate in infants but tenses the palate with anatomic maturation.

The nasopharynx is "a musculomembranous tube serving as a portal between the nasal chamber anteriorly and the oral pharynx inferiorly" (Handelman & Osborne, 1976, p. 243). The volume of the nasopharynx increases by about 80% from infancy to adulthood (Bergland, 1963). Tourne (1991), noting that the depth dimensions of the nasopharynx are established early in life, commented, "It is surprising to what a limited extent the anteroposterior diameter of the nasopharynx is increased" (p. 129). He explained this small size increment by pointing out that an acute angle of the cranial base produces a more vertical than horizontal direction of pharyngeal growth. The soft tissues of this region consist of the flexor neck muscles, superior pharyngeal constrictor muscles, fascia, mucous membranes, and lymphoid tissue. All of these vary with age, but lymphoid tissue has particularly large changes during development. It appears that the soft tissues of the nasopharynx gradually thicken until about 2 years of age (Capitanio & Kirkpatrick, 1970), and then remain virtually unchanged until about 15 years of age (Johannesson, 1968).

A particularly important feature of the nasopharynx in developmental studies is the lymphoid tissue known as the nasopharyngeal tonsil or "adenoids." This tissue is known to grow in early childhood and then atrophy. The status of the nasopharyngeal tonsil in infancy is not well known because the tissue is not radiographically visible until about the first year of

TABLE 2. Selected sources for anatomic development of the nasopharynx. Studies are listed chronologically. (*N* = number of subjects.)

Source	*N*	Age (years; mos)	Comments
King (1952)	550	0;3–16	X-ray cephalometry; number of subjects varied somewhat with different measures.
Subtelny (1957)	490	0–18	Measurements of length, thickness, and width of soft palate; measurements of vertical growth of nasopharynx.
Castelli, Ramirez, & Nasjleti (1973)	60	6–15	X-ray cephalometry; 20 subjects in each of 3 age groups.
Melsen (1975)	60	0–18	Histology and microradiography of autopsy specimens.
Handelman & Osborne (1976)	12	0;9–18	X-ray cephalometry.
Mazaheri, Krogman, Harding, Millard, & Mehta (1977)	15	0;6–6	Lateral cephalometry of velopharyngeal region; also data for 54 subjects with palatal clefts.
Jeans et al. (1981)	41	3–19	X-ray cephalometry at annual intervals
Scheerer & Lammert (1980)	80		Plaster model.
Melsen & Melsen (1982)	30 72 105	0–27 0–20 NR	Autopsy specimens; Autopsy specimens; dry skulls, various developmental stages. Extensive data on the palatomaxillary region.
Linder-Aronson & Leighton (1983)	46	3–16	X-ray cephalometry in a cross-sectional study of 28 boys and 28 girls.
Bjork & Skieller (1984)	14	4–18	X-ray cephalometry with metallic implant.
Lang & Baumeister (1984)	77	0–13	Measures of palatal width and height from dry skulls.
Bosma (1985)	—	—	Review article.
Coccaro & Coccaro (1987)	—	—	Review article with case studies.
Blocquel, Laude, Lafforgue, & Devillers (1990)	75	varies	X-ray cephalometry at two ages, separated by 1 to 7.5 years.
Chiba (1990)	40	varies	Dental casts taken within 2-year interval around emergence of first and second molars.
Kuehn & Kahane (1990)	10	Adult	Histology of soft palates from four men and six women.
Tourne (1991)	—	—	Review article.
Jeans et al. (1991)	41	3–19	X-ray cephalometry of the nasopharynx.
Gulisano et al. (1992)	50	0–16	MRI studies of nasopharynx. In Italian.
Kramer et al. (1992)	68	0;0–0;3	Dental casts made 48 hours after birth and at 3 mos.

life (Coccaro & Coccaro, 1987). Hypertrophy of the nasopharyngeal tonsil in childhood may obstruct the passageway for nasal air flow. It is of special interest to determine if the bony nasopharynx and lymphoid tissue have different rates of growth, given that disproportionate growth could constrict the air portal. Handelman and Osborne (1976) did indeed conclude that the rate of hypertrophy of the nasopharyngeal tonsil exceeded the growth increment of the nasopharynx during the preschool and early school years. The lymphoid hypertrophy compromised the nasopharngeal airway in many of the subjects in this age range. Jeans, Fernando, Maw, and Leighton (1981) confirmed this result, reporting that growth of the nasopharyngeal soft tissue exceeded that of the bony substrate during the period of 3 to 5 years. Similarly, Fujioka, Young, and Girdany (1979) reported that the nasal airway has its smallest size at age four. From these anatomic observations, it would be expected that velopharyngeal or nasal resistance to air flow would be increased in young children relative to older children or adults. Saito and Nishihata (1981) reported nasal resistance values that confirm this prediction. However, allowances should be made for individual differences in the growth of the pharyngeal lymphoid tissues (Pruzansky, 1975; Subtelny, 1954).

Subtelny's (1957) data indicate that the nasopharynx grows vertically about 12 mm from the first to the eighteenth year of life. Growth appeared to be quite uniform, a little less than 1 mm per year. Gulisano et al. (1992) used Magnetic Resonance Imaging to study nasopharyngeal anatomy in 50 subjects who ranged in age from 2 months to 16 years. Statistically significant differences were observed in both the bending radius and what the authors called a "shape factor" of the nasopharynx. Subtelny's (1957) measurements of the length of the soft palate indicate a rapid growth in the first 2 years of life, followed by a fairly uniform rate of growth until about 4 or 5 years. Thereafter, growth was slower. The annual increment is about 0.74 mm per year for both males and females. Maturity is not easily estimated from Subtelny's data, because a clear asymptote is not evident over the age range reported. Thickness of the soft palate increased rapidly during the first year of life but further increases were small. Maximum thickness was observed at about 14–16 years.

Velopharyngeal function for speech takes several forms, depending on sex, age, individual differences in velopharyngeal anatomy, and perhaps other unidentified factors. Sexual dimorphism of velopharyngeal valving in the adult has been described by McKerns and Bzoch (1970). In males the velum forms an acute angle in its orientation to the posterior pharyngeal wall, whereas in females the velum forms a right angle. A second difference is that the size of the area of seal is more extensive in females. Finally, the midpoint of closure in the nasopharynx is superior to the palatal plane in males but inferior to the plane in females. Possibly, these sex differences in valving pattern contribute to reported differences in aerodynamic function (Thompson & Hixon, 1979). It would be interesting to know if similar sex differences are present in children, and if they are, at what age they emerge.

Marked individual differences have been observed in the pattern of velopharyngeal valving (Croft, Shprintzen, & Rakoff, 1981; Shprintzen, 1992; Skolnick, McCall, & Barnes, 1973). The four most frequently observed patterns are coronal (accomplished mainly by anteroposterior movements of the velum), sagittal (accomplished primarily by lateral pharyngeal wall movement), circular (performed by essentially equal movements of the velum and lateral pharyngeal walls), and circular with Passavant's ridge (same as the circular pattern but aided by an anterior movement of the posterior pharyngeal wall). These different patterns would seem to allow for substantial compensation for various configurations of the nasopharyngeal anatomy. Relatively little is known about the development of these patterns of velopharyngeal valving. However, it is widely acknowledged that velopharyngeal valving in very young children is predominantly velar-adenoidal (Croft, Shprintzen, & Ruben, 1981; Skolnick, Shprintzen, McCall, & Rakoff, 1975). Siegel-Sadewitz and Shprintzen (1986) observed changes in velopharyngeal valving in 60% of normal subjects in a prepubertal-postpubertal comparison. It appears likely, then, that nearly all children pass from a velar-adenoidal closure pattern to another pattern, and a majority may then change again before adulthood. Possibly, these developmental changes in valving pattern are related to the observation of Thompson and Hixon (1979) that anticipatory coarticulation of nasalization increases with age.

Summary

The pharynx grows primarily in the vertical dimension, roughly tripling its length from birth to adulthood. The development of the nasopharynx has several important features, including (a) a substantial remodeling during infancy which changes the actions of some palatal muscles, (b) relatively rapid growth of the soft palate in the first 2 years of life with more gradual growth thereafter, (c) hypertrophy of the nasopharyngeal tonsil which narrows the nasal passageway in young children, and (d) a likely change in pattern of velopharyngeal valving during development in most subjects.

Clinical Note:

Pharynx and Velopharynx

An elongated uvula has been associated with respiratory distress, apnea, and laryngospasms, apparently caused by contact of the uvula with the vocal folds and supraglottic structures (Shott & Cunningham, 1992). Infants from 1–3 months of age have the highest incidence of laryngospasm in pediatric populations (Roy & Lerman, 1988). This high incidence is understandable, given the tendency for palato-laryngeal engagement in the neonate.

A substantial literature surrounds the development and clinical implications of the lymphoid tissue in the oropharynx, laryngopharynx, and nasopharynx. No attempt will be made to review this literature here, but the interested reader might begin with the articles of Behlfelt and colleagues describing the complex relations between enlarged tonsils and head posture, dentition, and craniofacial morphology (Behlfelt, Linder-Aronson, & Neander, 1990; Behlfelt, Linder-Aronson, McWilliam, Neander, & Laage-Hellman, 1989). Related work by Hultcrantz et al. (1991) also demonstrates the important relation between tonsillar obstruction and craniofacial growth.

The predominant pattern of velopharyngeal valving in young children appears to be velar-adenoidal (Croft et al., 1981; Skolnick et al., 1975). With atrophy of the adenoid (pharyngeal tonsil), a more mature pattern of valving comes into place. However, a change in some aspects of velopharyngeal valving between prepuberty and postpuberty seems very common (Siegel-Sadewitz & Shprintzen, 1986). Four major patterns of velopharyngeal valving have been identified, as reviewed earlier in this section. Developmental changes in aerodynamic aspects of velopharyngeal function have been described by Thompson and Hixon (1979). New methods of acoustic analysis have been reported by Slawinski and Dubanowicz-Kossowska (1993), who described a procedure for the assessment of hypertrophy of the nasopharyngeal tonsil (adenoid). Such noninvasive methods are particularly valuable to record longitudinal changes in velopharyngeal function. Nasoendoscopic and aerodynamic evaluations can be used successfully with children as young as 2 years of age if suitable materials and methods are applied (Lotz, D'Antonio, Chait, & Netsell, 1993). The availability of these methods increases the need for quantitative data on normal pediatric populations. The normal pattern is that values of nasal resistance are relatively high during early childhood, about 3 to 8 years, but subsequently decline into adulthood (Saito & Nishihata, 1981; Warren, Duany, & Fischer, 1969).

Lips and Labio-Oral Region

Selected primary sources of information on the lips and labio-oral tissues are summarized in Table 3.

The lips change considerably during the first 2 years of life, adjusting for their biologic function of feeding. Aside from drastic changes in both the size and shape of lips, the epithelial composition changes. Thach (1973) divided the lips into three zones: hairy cutaneous, glabrous, and papillary. He reported the papillary zone, which appears flushed in term infants, to have a higher percentage of papillae per mm (12–15 papillae) rendering the surface mobile and adhesive, which facilitates oral seal during suckle feeding.

According to Burke (1980) the lips continue to grow in width, height, and convexity into the mid to late teens, but the most drastic reconfiguration occurs during the first 2 years of life when lip width almost doubles (20 mm to 39.4 mm) as lip convexity and height decrease. However, after the age of 4 years, all three dimensions increase gradually, with smaller growth changes noted for girls than for boys (even when there is no difference in height and weight). Overall, these findings are consistent with the findings of Farkas, Posnick, Hreczko, and Pron (1992); Vig and Cohen (1979); Fasika (1993); as well as Mamandras (1984, 1988). Burke (1980) further reported that typically, but not consistently, the lower lip makes more forward growth than the upper lip. Madzharov and Madzharova (1992), in a study of 2,300 Bulgarians, concluded that the upper lip has its largest growth in the preschool years but continues to grow in size until the age of 80 years.

In general agreement with Burke's (1980) findings, Fasika (1993), who studied Nigerian children, reported steady growth in the lips (mouth width and length of the upper vermilion arc) throughout the ages he studied (birth to 12). Fasika further described the height of the cutaneous upper lip and the height of the upper vermilion border to have rapid growth up to age 6, following which growth slows. The latter finding is consistent with the conclusion of Farkas, Posnick, Hreczko, and Pron (1992) that the upper labio-oral region appears to reach maturity between the ages 6 and 11.

Vig and Cohen (1979), who examined upper and lower lip length (which is grossly equivalent to Mamandras' lip length measurements) over the age range of 4 to 20 years, reported the most rapid growth between 10–17 years. Similarly, Mamandras (1984), who examined maxillary and mandibular lip length, thickness, and area over the age range of 8–18, described rapid growth between 10–16 years, with significant gender differences as well as differences between the maxillary and mandibular lips. Between the ages 8–18, the average maxillary lip length increases approximately 17% (males 21%, females 12%), whereas the mandibular lip length increases 29% (males 39%, females 19%). Growth spurts in maxillary lip length and thickness were noted between the ages of 10–16 and 8–16 years, respectively, for males, and 10–14 and 12–14 years, respectively, for females. For mandibular lip length and thickness, growth spurts were noted between the ages 12–16 and 10–16

TABLE 3. Sources for developmental anatomy of the lips. US = young adults of unspecified ages. Number of subjects (*N*) and their ages (years) are given for each paper. Studies are listed chronologically.

Source	*N*	Age (*in years*)	*Comments*
Vig & Cohen (1979)	50	4–20	Lateral x-ray cephalometry.
Burke (1980)	12 1	9–18 0–7	Stereoscopic photography used to produce contour maps of face; subjects include six pairs of twins and one child studied from 3 weeks to 7 years.
Bishara et al. (1984)	35	5–25.5	Lateral cephalometry in a study of facial dimensions.
Mamandras (1984)	28	8–18	Lateral x-ray cephalometry at 2-year intervals.
Farkas et al. (1984)	89 100	18–25 US	Measurement of vertical profile of face; vermilion arcs and mouth width.
Mamandras (1988)	32	8–18	Lateral x-ray cephalometry at 2-year intervals.
Nanda et al. (1990)	40	7–18	Annual x-ray cephalometry; longitudinal growth changes in soft tissue profile.
Madzharov & Madzharova (1992)	2,300	0–102	Direct cephalometry of upper lip.
Farkas, Posnick, Hreczko, & Pron (1992)	1,593	1–18	Direct cephalometry of naso-labial region of white North Americans.
Zylinski, Nanda, & Kapila (1992)	60	5–32	Lateral x-ray cephalography focusing on soft tissue profile in caucasian males.
Fasika (1993)	240	0–12	Direct cephalometry of Nigerian children.

years, respectively, for males and 10–16 and 10–12 years (also 14–16), respectively, for females. Mamandras (1984) reported that the maxillary lip area was significantly greater for males between the ages of 10–18 years. In contrast, the mandibular lip area was significantly greater for females at the age of 12; however, this relation is reversed by age 18. Figures 12 and 13 summarize Mamandras' measurements of lip area for the maxillary and mandibular lips, respectively.

As an example of CMA studies of the mandible and mandibular lip, a comparison of various measures of mandibular lip length and mandibular depth is given in Figures 14 and 15, respectively. Despite differences in measurement procedures, overall agreement in derived values is reasonably good.

Summary

The development of the lips, despite racial differences in absolute measurements, can be described in terms of an early growth spurt between birth and 2 years and a later spurt occurring within the range of 10–17 years. The latter acceleration in growth appears to span growth acceleration of the mandible in males (Walker & Kowalski, 1972). Labial growth may continue into adulthood, at least in males.

Clinical Note:

Perioral Muscle Function

Jansen et al. (1990) described simple bedside methods to assess perioral muscle function, using measures of lip length and snout (anterior projection of the oral region) as indices of the degree to which mouth width can be lengthened or shortened. Patients with facial muscle weakness were observed to have low values of both indices. Unfortunately, comparable data do not seem to be available for children. The snout/lip-length measures are easily derived and may have potential for the quantitative evaluation of perioral functional development in children. These measures could be particularly valuable in evaluating changes in motor control of the labial muscles, for example, in children with cerebral palsy or labial trauma.

Less quantitative procedures also appear to be useful in the assessment of nonspeech motor behaviors related to speech articulation in children (Qvarnstrom, Jaroma & Laine, 1993, 1994). Although questions remain about the relationship between speech and nonspeech motor control of the oral musculature, improved methods of examination hold promise to illuminate this issue.

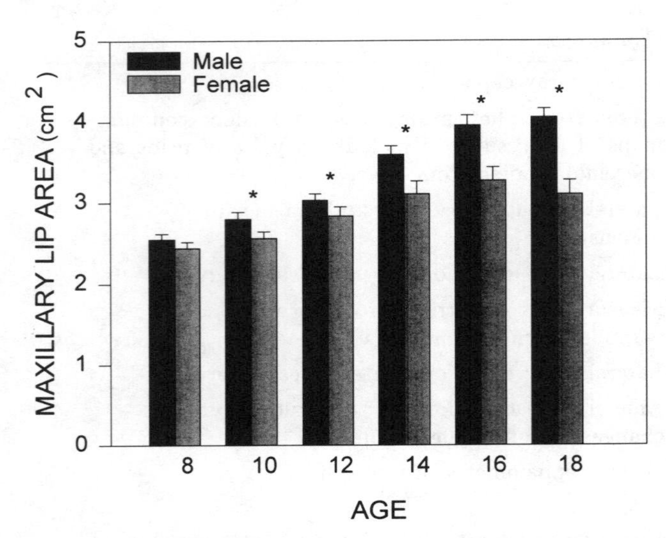

Figure 12. Maxillary lip area (means and standard error bars) as a function of age for males and females. Asterisk (*) indicates a significant gender difference. Drawn from measurements reported by Mamandras (1984). Used with permission.

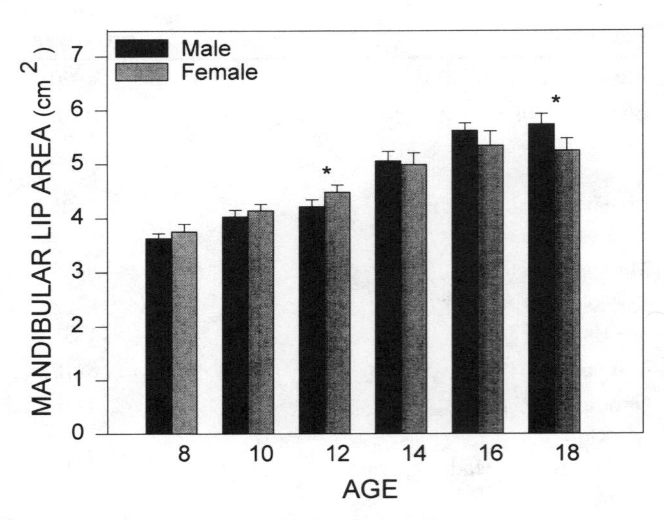

Figure 13. Mandibular lip area (means and standard error bars) as a function of age for males and females. Asterisk (*) indicates a significant gender difference. Drawn with permission from measurements reported by Mamandras (1984).

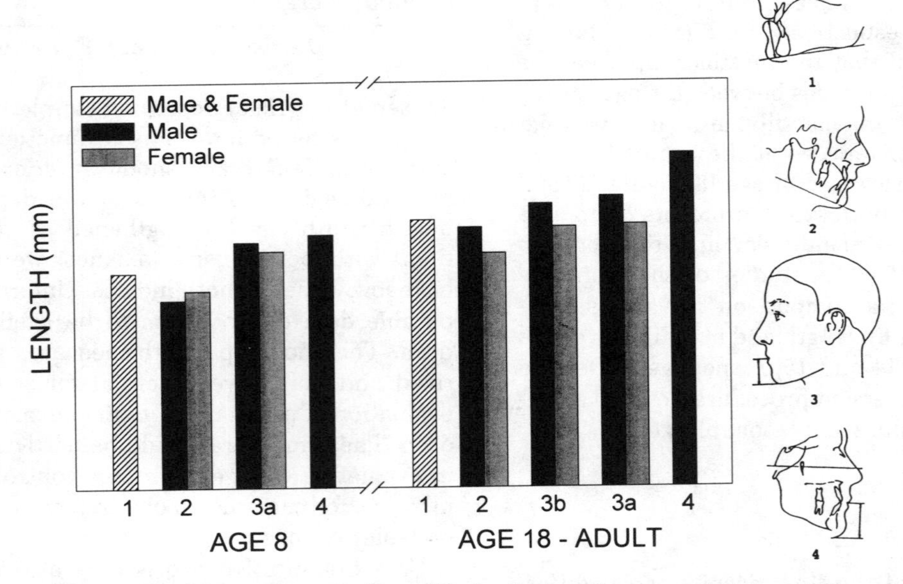

Figure 14. Measurements of mandibular lip length reported in four different studies: 1. Vig and Cohen (1979); 2. Mamandras (1988); 3a. Farkas, Posnick, Hreczko, and Pron (1992); 3b. Farkas et al. (1984); 4. Zylinski et al. (1992).

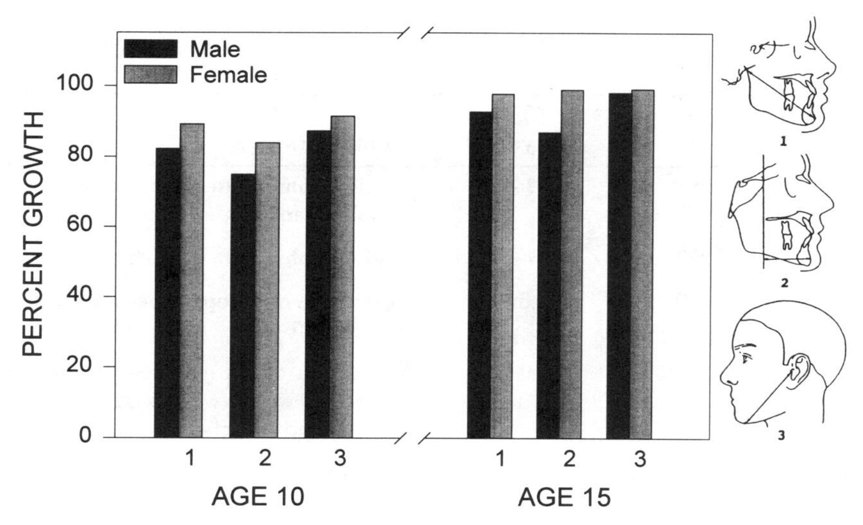

Figure 15. Measurements of mandibular depth reported in three different studies: 1. Bishara et al. (1984); 2. Nanda et al. (1990); 3. Farkas, Posnick, and Hreczko (1992b).

Tongue

Selected primary sources of information on the tongue are summarized in Table 4.

Few studies have been published on the development of the tongue, and the few that have been published are rather difficult to integrate into a cohesive developmental account because of differences in procedure, limitations in data for some ages, and differences in the subject indices by which measurements are reported. For example, both Wittman (1977) and Siebert (1985) reported that tongue weight correlates closely with crown-to-heel height. Siebert obtained a Pearson's correlation coefficient of $r = 0.928$ ($p < .001$). But it should be noted that this statistic was based on measurements of normal tongue size from 44 prenatal and 39 postnatal subjects, so that the prenatal specimens could have strongly influenced the correlation. Wittman (1977) presented ratio data for tongue weight relative to both body weight and body height. Apparently, she viewed the proportions as evidence for a relatively constant ratio of tongue weight to body length. Wittman did not report a statistical test of significance. A Pearson's correlation coefficient of $r = .504$ ($p > .05$) was computed by the present authors from the data for tongue weight and body length in her article. The low correlation obtained from Wittman's data may reflect the fact that her measurements were from adult tongues, whereas Siebert's measurements were from a combined sample of pre- and postnatal specimens. Both Siebert (1985) and Wittman (1977) concluded that subject height is the preferred reference for tongue measurements. However, Brulin and Talmant (1976) and Talmant and Brulin (1976) used chronological age as their reference. For comparison purposes, we have utilized the 50th percentile body length data from the Department of Health (in Kent, 1994) to determine the corresponding chronologic age. This index appears to be appropriate because the ages thus determined correspond with Siebert's subjects' chronological age range.

The tongue of the newborn occupies almost the entire oral cavity. Its surface comes into contact with the gums laterally and with the roof of the mouth superiorly. Crelin (1973) reported that, although all the types of adult lingual papillae, including taste buds, are present before birth, the newborn infant has an increased suckling response only to sweet taste. Crelin (1973) described the newborn tongue to be about 4 cm long, 2.5 cm wide, and 1 cm thick. Siebert's (1985) measurements of the newborn tongue are roughly equivalent (4.3 cm long, 2.9 cm wide, 1.2 cm thick, and 9.4 g in weight). Laitman and Crelin (1976) and Crelin (1973) reported that, during the first year of life, the posterior third of the tongue begins to descend into the neck to become part of the anterior wall of the pharnx. This descent is reportedly completed at approximately age 4 or 5 years. Siebert's (1985) developmental study of tongue size determined tongue measurements by age 4 to 5 to be 5.8 cm long, 3.9 cm wide, 1.6 cm thick, and 23.3 g in weight.

The classic account of tongue growth (Scammon, 1930) is that the tongue reaches approximately 90%

TABLE 4. Sources on anatomic development of the tongue. *N* = number of subjects; all ages in years. Studies are listed chronologically.

Source	*N*	Ages (in years)	Comments
Talmant & Brulin (1976)	288	7–17	Radiographic data for a correlative study of the sagittal development of the face and tongue. In French.
Brulin & Talmant (1976)	292	7–17	Radiographic data on lingual growth. In French.
Wittmann (1977)	20	Adult	Dissection of autopsy specimens, 10 males and 10 females, ages 44–85.
Siebert (1985)	83	Fetal–10.5	Dissection of autopsy specimens. Data on tongue length, thickness, width, weight. Most information is reported on fetuses and infants.
Lauder & Muhl (1991)	19	Adults	Magnetic resonance imaging used to estimate volume of tongue, oropharynx, and oral cavity.
Sato & Sato (1992)	18	Fetuses, adults	Microscopy (light, scanning electron, transmission electron) of myofibrils, mitochondria, and other cytoplasmic organella.

of its adult size by the age of about 8–10 years. If so, then the tongue is similar to the brain in reaching a large proportion of its final size in early childhood. However, more recent data do not confirm this early maturation of lingual size. Siebert's (1985) data on tongue size, involving 83 autopsy specimens (44 neonatal specimens and 39 specimens ranging in age from birth to 10.5 years, but with unequal number of specimens per age group), showed that from birth to puberty the tongue doubles in length, width, and thickness and triples in weight. In comparing his data to adult data from Wittmann (1977), Siebert concluded that tongue weight increases tenfold from birth to adulthood (Figure 16). Siebert believed that his dissection was similar to that used by Wittman and therefore assumed that a direct comparison of the data from the two studies was permissible. He furthermore speculated that the tongue may continue to grow into adulthood, possibly demonstrating a lifespan hypertrophy. Brulin and Talmant (1976) and Talmant and Brulin (1976) reported tongue measurements in a sample of 288 subjects aged 7–18 years. They determined the sagittal lingual surface from tracings of lateral radiographs. Their data show growth spurts at the relatively discrete age intervals of 9–11 and 13–14 years, which they compared to Woodside's (Woodside, 1969; cited by Talmant & Brulin, 1976) findings on mandibular growth spurts. Furthermore, they determined that lingual maturity is reached at about age 16. The conclusion that lingual growth continues until puberty also was reached by Kerr et al. (1991), who measured the tongue dorsum area in 57 subjects aged 9.3 to 19 years. They establish the age of 15 years as the age of lingual maturation. These observations do not support Siebert's hypothesis of a lifespan lingual hypertrophy. However, they are consistent with the general maturational schedule of the craniofacial complex described earlier for the mandible and lips.

Ardran and Kemp (1972) speculated that the tongue may increase in size in partial edentia, but Wittmann (1977) did not confirm this hypothesis in her sample of 20 adults with various dentitions (e.g., normal, carious, few upper or lower teeth, partially edentulous, and edentulous). Wittmann reported the average male tongue to weigh 97.3 g (range 74–110 g), and the average female tongue to weigh 80 g (range 55–108 g). Thus, the male tongue is approximately 20% heavier than the female tongue. Lauder and Muhl (1991) also reported gender differences in adult tongue size and found that tongue volumes in adults were correlated with subject body weight ($r = 0.86$ for coronal and $r = 0.82$ for sagittal orientations). The Lauder and Muhl article is also important as a guide to the use of magnetic resonance imaging in estimating tongue volume.

As Siebert (1985) pointed out, an understanding of tongue growth and the relation of tongue size to the other structures in the oropharynx, such as the dental arch, is relevant to developmental studies. Also, this information is needed to recognize and interpret abnormal conditions because both micro- and macroglossia occur in a number of diseases (Ardran & Kemp, 1972; Siebert, 1985; Wittmann, 1977). As noted

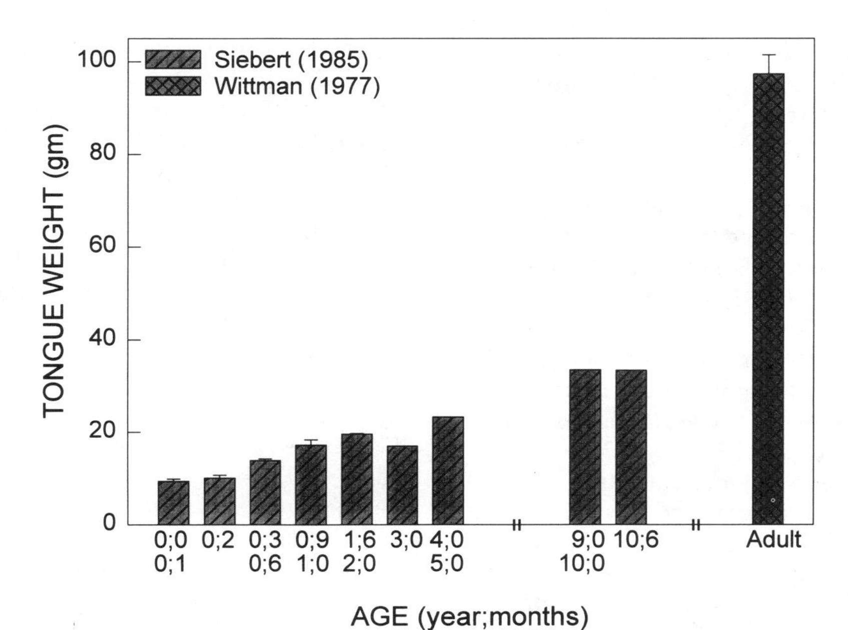

Figure 16. Weight of the tongue at different ages, based on measurements reported by Siebert (1985) and Wittman (1987). The standard error bars were calculated from data given in the original articles.

above, Wittmann (1977) reported gender differences in tongue weight. Tamari, Murakami, and Takahama (1991) obtained evidence of tongue volume differences between males and females. However, to our knowledge, whether such differences are present at birth or reflect developmental changes has not been reported in the literature. Data on racial differences also are limited, but there are indications that such differences exist. For example, racial differences have been observed in measurements of maxillary arch width (Gross, Kellum, Franz et al., 1994; Gross, Kellum, Michas et al., 1994). Assuming that tongue size is correlated with maxillary arch width, the racial differences reported for the latter suggest that racial differences in tongue size also may exist.

Summary

The tongue of the newborn initially occupies most of the oral cavity. During the first 5 years of development, the posterior part of the tongue gradually lengthens into the pharyngeal cavity to create additional oropharyngeal space. Regarding its growth trend, the tentative conclusion is that the tongue increases in all its dimensions and reaches its adult size at about age 16, which parallels the general maturational schedule of the craniofacial complex. The classic report that the tongue reaches nearly adult size in early childhood has not been confirmed by recent observations. Although Siebert (1985) speculated that the tongue may continue to grow into adulthood, it seems that lingual maturation is achieved at about 16 years following circumpuberal growth spurts. The growth of the tongue is in general harmony with that of the mandible and lips. Tongue size appears to correlate well with body weight or height. These conclusions should be regarded as tentative, given the very few developmental studies of the tongue.

Clinical Note:

Tongue Size

Tamari, Murakami, and Takahama (1991) described procedures by which tongue volume can be estimated from the stretched length of the tongue. Their paper provides a foundation for the clinical evaluation of tongue size, using relatively simple methods. On the more technical side, estimation of tongue volume using magnetic resonance imaging is described by Lauder and Muhl (1991). The tongue appears to grow in general harmony with the mandible and lips. A marked discrepancy in the relative growth of these structures may therefore signal anomalous development.

Ardran and Kemp (1972) reported that patients with a small tongue have difficulty clearing the bolus of food from the mouth due to difficulty apposing the tongue to the hard palate. On the other hand, patients with a large tongue have difficulty clearing the forepart of their mouth of saliva. Also, they may have difficulty in using the tongue to make consonants. These authors speculated that increases in tongue size in older adults may be related to partial edentia. An enlarged tongue also has been suggested as an agent in the pathogenesis of airway obstruction in sudden infant death syndrome (SIDS) (Tonkin, 1975; cited by Siebert & Haas, 1988). To test this hypothesis, Siebert and Haas (1988) obtained weight and size measurements of the tongues of 100 infants who had died of SIDS and 40 control infants matched for age and body size. The tongues of the infants with SIDS were on the average 2.1 grams heavier than those of the control infants.

Hale et al. (1992) reported an association between trisyllable diadochokinetic rates and dentalized resting tongue posture and dentalized swallow patterns. This association was interpreted to reflect poorer motor control in children who maintain a forward tongue rest and swallow postures. Lingual motor control was judged to improve between two age groups of subjects (6–11 years and 19–23 years) in a study of toffee clearance by Speirs and Maktabi (1990). Relative tongue placement in children also was considered in a study by Qvarnstrom et al. (1993). These authors concluded that anterior placements of the tongue were associated with immaturity of speech movements and a high likelihood of spontaneous recovery. Posterior placements of the tongue were thought to be related to a difficulty in coordinating movements of different tongue muscles.

DEVELOPMENT OF THE LARYNX

Selected primary sources of information on the larynx are summarized in Table 5. Laryngeal development is discussed under the headings of the epiglottis, trachea, and larynx, with the larynx subdivided into the thyroid cartilage, cricoid cartilage, arytenoid cartilage, vocal folds, and glottis. An overall summary is given at the end of the section on the larynx.

Epiglottis

The epiglottis is a leaf-shaped elastic cartilage hanging over the entrance to the larynx. It is attached superiorly to the hyoid bone via the hyoepiglottic ligament and inferiorly to the thyroid cartilage via the thyroepiglottic ligament. The lower lateral margins of the epiglottis are attached to the arytenoid apexes through the aryepiglottic folds. Fried, Kelly, and Strome (1982) took note of the fullness of the infant mucosal folds by describing their thick epiglottis, aryepiglottic folds, and the tissue overlying the arytenoids. As will be noted later, the infant arytenoid cartilages are proportionately large in size, and the thick aryepiglottic folds have a medial bowing which gives the thick epiglottis an "omega shaped" appearance when viewed from above (Fried et al., 1982).

Crelin (1973) reported the vertical length of the newborn epiglottis to be 1.2 cm. During development, the epiglottis undergoes a number of changes. At the microscopic level, Adam and Pohunkova's (1986) study showed that the development of the elastic cartilage of the human epiglottis continues postnatally, reaching maturity at puberty (female specimen, age 17). However, they also reported structural changes in adulthood. They characterized postnatal changes to be chiefly quantitative (growth of chondrocytes and an increase in the amount of intercellular matrix), whereas changes in adulthood are qualitative (changes in the distribution of chondrocytes and fragmentation of the elastic fibers). Such changes may be related to Bosma's (1975b) report that during development the epiglottis enlarges and becomes firmer.

There is no general agreement as to the function of the epiglottis. However, in infants, the epiglottis is believed to play an important role in separating the upper respiratory tract from the upper digestive tract (Crelin, 1973; Gopal & Gerber, 1992; Laitman & Crelin, 1976). The infant epiglottis is positioned very high in the oropharyngeal tract such that it is in contact with the velum (Crelin, 1973, Gopal & Gerber, 1992, Laitman & Crelin, 1976). This vocal tract configuration allows the infant to carry out the functions of breathing and feeding simultaneously. The cineradiographic findings of Sasaki, Levine, Laitman, and Crelin (1977) show that, up to 2 months of age, the epiglottis is in contact with the soft palate, rendering the infant to be an obligate nasal breather. However, at 4–6 months, this contact is present primarily during deglutition, reflecting a transitional period from nasal to oral tidal breathing.

Crelin (1973) stated that, at rest, the infant epiglottis is at the level of the second or third cervical vertebra, whereas the adult epiglottis is at the lower third or fourth cervical vertebra (the fifth, according to Gopal & Gerber, 1992). Crelin (1973) reported that, during deglutition, the infant epiglottis is elevated to the level of the first cervical vertebra and makes direct contact with the soft palate. Alternatively, findings from cineradiograms (Sasaki et al., 1977) show that, during respiration, the tip of the epiglottis is at the level of the first cervical vertebra in neonates and infants up to age 4 months. However, by age 6 months, it descends to the level of the third cervical vertebra.

TABLE 5. Sources on anatomic development of the larynx. *N* = number of subjects; NR = not reported. Studies are listed chronologically.

Source	*N*	Ages (in years)	Comments
Holibkova (1973)	—	0–95	Thickness of lymphatic tissue in larynx.
Too-Chung & Green (1974)	67	0–15	Growth of cricoid cartilage.
Tucker et al. (1976)	4	0–73	Electron microscopy of laryngeal epithelium in a newborn, a 51-year-old woman, a 60-year-old man, and a 73-year-old man.
Tucker & Tucker (1979)	—	—	Review article.
Kahane (1978)	20	9–19	Autopsy specimens
Kazarian et al. (1978)	324	0–20	Data on adults up to 65 years also reported. Data for both males and females but unequal numbers for some ages. In Russian.
Hirano et al. (1981)	38	0–18	Autopsy specimens; also includes data on 50 adults.
Wright, Ardan, & Stell (1981)	45	0;11–13;2	Head and neck radiography.
Fried et al. (1982)	—	—	Review article; good summary and helpful illustrations.
Kahane (1982)	20	9–19	Autopsy specimens.
Kahane (1983)	12	30–90	Autopsy specimens, one male and one female from each decade. Histological study, survey of changes with aging of connective tissue.
Kahane & Kahn (1984)	9	0;1–0;6	Weight measurements of intrinsic laryngeal muscles.
Bosma (1985)	—	—	Review article.
Verhulst (1987)	—	—	Brief review article. In French.
Kahane (1988)	40 24 16	0;1.5–92 20–90	Autopsy specimens, growth of cricoarytenoid joint. Stereomicroscopy. Histologic studies.
Bonnaure-Mallet & Lescoat (1989)	9 6	45–70 Fetuses (22–38 wks)	Light and electron miscroscopy of the vocal folds
Sato et al. (1990)	37	Newborn to Adult	Study of the distribution of elastic cartilage in the arytenoids. Results on 10 newborns and 27 adults.
Sellars & Keen (1990)	26	0;0.6–3	Growth of the cricoid ring.
Sakai et al. (1990)	62	19–78	MRI of the larynx; two normal excised larynges also studied.
Cohen et al. (1992)	NR	0–6;10	Development of collagen.
Garel, Contencin, Polonovski, Hassan, & Narcy (1992)	40	0–15	General description of sonographic imaging of laryngeal structures; no quantitative data.
Hirano & Sato (1993)	2	Newborn, Adult	Histologic color atlas of the larynx, based on cadavers (one newborn female and one 27-year-old male).
Claassen & Kirsch (1994)	19	Fetal–74	Immunofluorescence staining of thyroid cartilage to localize fibrillar collagen types I and II.

Trachea

The larynx is connected to the superior part of the trachea and to the pharynx inferior to the tongue and hyoid bone. The trachea in adults has a length of about 11 to 12 cm and a diameter of between 2 and 2.5 cm (Zemlin, 1988). As a rough rule of thumb for older children, the tracheal diameter in mm corresponds with the child's age in years. According to Tucker and Tucker (1979), the tracheal diameter varies from 3 mm in premature infants to 25 mm in larger adults. Tracheal rings are composed of hyaline cartilage about 2 mm thick, and there are approximately 4 rings per 10 mm of tracheal length. The total number of rings varies from 16 to 20 (Tucker & Tucker, 1979; Zemlin, 1988). The first (most superior) tracheal ring is the largest and connects with the cricoid cartilage through the cricotracheal ligaments.

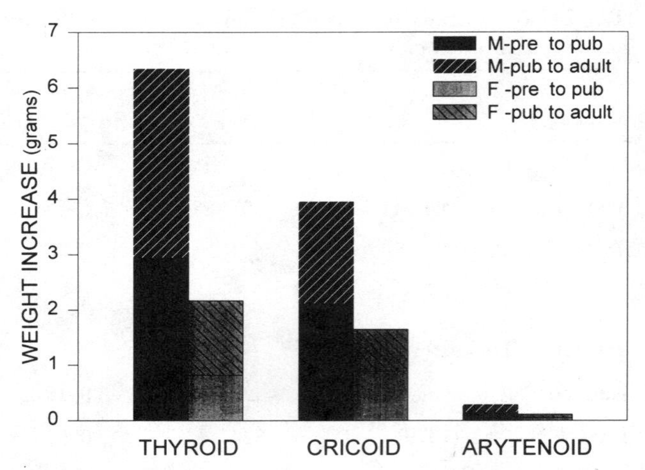

Figure 17. Weight increases of the thyroid, cricoid, and arytenoid cartilages from prepuberty to adulthood. Based on measurements reported by Kahane (1978, 1982). Used with permission.

Larynx

Laryngeal growth is reported to be extremely active during the first 18 months of postnatal development, with the laryngeal cartilages, muscles, mucous membranes, and submucosal tissues becoming firmer and less pliable (Tucker & Tucker, 1979). The larynx of the infant is commonly estimated to be one third the dimension of the adult larynx (Bosma, 1985; Tucker & Tucker, 1979; Verhulst, 1987). According to Crelin (1973), the infant's larynx is 2 cm long and has a comparable width at the level of the upper margin of the thyroid cartilage. Similarly, Tucker and Tucker (1979) noted infant laryngeal width to be 2 cm. Crelin (1973) also commented on an early sexual dimorphism, with the larynx being larger and longer in males than in females by the age of 3 years. Kahane (1978, 1982) provided measurements of larynges and described growth of the larynx from prepuberty to adulthood. His results are shown in Figure 17 for measurements of the weight of the laryngeal cartilages. He reported morphologic congruence in the male and female prepubertal larynx with a clear sexual dimorphism developing by puberty. Kahane's measurements showed quantitatively more growth in male laryngeal cartilages and male vocal folds than in female structures.

Kahane (1978, 1982) characterized the growth of the laryngeal cartilages as an increase in size with maintenance of their basic shape, with the exception of the anterior aspect of the male thyroid cartilage. Furthermore, his findings showed increases in size of the laryngeal cartilages to be accompanied by significant increases in weight, with male cartilages weighing 2–3 times more than female cartilages. Kahane estimated that one half of the overall increase in cartilage weight is attributable to an increase in size, with the remainder due to calcification and ossification of the cartilages.

Data on the weights of the intrinsic muscles of the larynx were reported by Kahane and Kahn (1984) in an infant-adult comparison. The muscles studied were the cricothyroid (CT), posterior cricoarytenoid (PCA), interarytenoid (IA), and the lateral cricoarytenoid (LCA)/thyroarytenoid (TA) combination. (The LCA and TA were weighed together as a unit because of the difficulty of separating their fibers.) The weights of these muscles in the infant, expressed both in mg and as a percentage of the values in the adult were: CT — 74 mg, 10.6%; PCA — 45 mg, 11%; IA — 29 mg, 9.6%; LCA/TA — 76 mg, 14%. It was concluded that the infant's intrinsic muscles are proportional in weight to their values in the adult. The cricothyroid was the largest individual muscle by weight. Kahane and Kahn commented, "On the basis of weight alone, the cricothyroid muscle appears to hold a preeminent position in the hierarchy of infant laryngeal muscles" (p. 132). The relatively large size of the CT was thought to reflect the need for a muscle to adjust the length and tension of the vocal folds during vocalization behaviors in infancy.

There are some discrepancies in the descriptions of laryngeal location or height in the neck, and also in measurements of the laryngeal structures. Such differences may be due to the age at which measurements were made, the crown-heel height and weight of the subjects, measurement procedures used, and preservation used for cadaver specimens. In addition, such discrepancies may reflect racial differences. The latter factor may be in part related to reports that crown-heel length is a better predictor of laryngeal growth than chronologic age (Kahane, 1982).

In the 5- or 6-week-old embryo, the larynx lies at the level of the basi-occiput, but it descends considerably by birth (Negus, 1962). The infant larynx is positioned high in the neck and is in close approximation to the hyoid bone (Fried et al., 1982; Laitman & Crelin, 1976; Tucker & Tucker, 1979; Verhulst, 1987). Bench (1963) considered the larynx at birth to be located between the third and fourth cervical vertebrae. Gopal and Gerber (1992) contrasted the infant's laryngeal position of approximately the level of the second cervical vertebra with the adult's position of close to the fifth cervical vertebra. According to Wind (1970), laryngeal position drops from the level of the third to the sixth cervical vertebra between birth and the age of 20 years. Wind also notes that the descent continues in adulthood, but at a much more gradual rate, so that laryngeal position is at the level of the seventh cervical vertebra at the age of 80 years. The high position of the larynx in infants serves to protect the airway but may restrict its phonetic repertoire (Crelin, 1973). As the larynx descends, its function evolves to meet phonatory requirements and provide structural support (Fried et al., 1982). The descent of the larynx begins around 4 to 6 months postnatally to produce the sharp bending of the vocal tract in the human adult. By age four, the elevated larynx assumed during deglutition can no longer reach the soft palate.

Differences in the position of the larynx between infants and adults are highlighted in the medical literature because intubation difficulties and subsequent subglottic stenosis may arise if such position differences are not taken into account (Fried et al., 1982; Tucker, 1980). The descent of the larynx also relates to positioning during feeding. In early life the infant typically is fed in a supine position or at an incline, but by 7 months an upright feeding position is common.

Although the focus of the present review is on postnatal changes in the location, shape, size, and function of the developing larynx, it should be emphasized that such developmental changes are accompanied by microanatomic changes. Cohen et al. (1992) stated that histologically the "cartilages of the larynx have been determined to be hyaline, except for the epiglottis, which is elastic fibrocartilage" (p. 328). However, Kahane (1983) and Sato, Kurita, Hirano, and Kiyokawa (1990) reported that the thyroid and cricoid are composed of hyaline cartilage, the epiglottis of elastic cartilage, and the arytenoid of both hyaline and elastic cartilage. Kahane and Sato et al. noted that, unlike the hyaline cartilages of the larynx which ossify with age, the elastic cartilages never ossify. Kahane estimated the onset of ossification to be in the third decade for males and the fourth for females. But according to Claassen and Kirsch (1994), mineralization and ossification begin shortly after the larynx reaches its final size at about 15–20 years of age. They also reported that the female larynx never completely ossifies.

Cohen et al. (1992) studied the biochemical makeup of the fibrous connective tissues that comprise laryngeal cartilages, specifically the collagen (protein) content in the epiglottis, thyroid, cricoid, and arytenoid. Their observations were based on six postnatal specimens aged from 1 day to 70 months. Cohen et al. established the different types of collagen that change dynamically during laryngeal growth and development. An acceleration of these changes was noted in the first month of life. For example, during this early period, there was an increased attachment (cross-linking) of Type I collagen, the most prevalent type of fibrous connective tissue in vertebrates. Cohen et al. commented that this maturational process contributes to the stability and the mechanical properties of the collagenous tissues. Also, they observed the emergence of Type II collagen (cartilaginous) which was absent at birth from all of the laryngeal cartilages they examined. Thus, changes in the fibrous connective tissue contribute to the structural framework of the larynx and assist in the function of its components. It appears that such changes continue into adulthood. For example, Claassen and Kirsch (1994) analyzed the distribution of collagen in the thyroid cartilage of various ages (prenatal to late adulthood). They described matrix changes throughout the lifespan with mineralization and ossification occurring earlier in the male than in the female thyroid cartilage.

Tucker, Vidic, Tucker, and Stead (1976) reported on developmental changes in the laryngeal epithelium that reflect the biologic integration of form and function. The most striking developmental difference in the cellular composition of the laryngeal epithelium was in the distribution of ciliated cells (the function of which is related to secretory activity and transport). The infantile epithelium is reportedly restricted in its ciliary area, and the cilia are short and thin.

Bosma (1985) and Kersing (1986; cited in Verhulst, 1987) reported on structural changes in laryngeal musculature. Verhulst (1987) summarized Kersing's finding on two types of muscle fibers. Type I muscle fibers are characterized by long prolonged contraction, whereas Type II muscle fibers have a short and fast contraction. The newborn infant has more Type II than Type I muscles. The predominance of Type II muscles allows the infant larynx to act like a sphincter with an important spasmodic and quick closure. Hence, infant vocalization is brief and unvaried. Gradually, Type I muscles develop allowing increased control of vocal modifications and prolonged vocal expression.

Recent attempts to use magnetic resonance imaging of the larynx indicate that this technique has excellent potential to reveal anatomic details of the major cartilages, extrinsic and intrinsic laryngeal muscles, ligaments, and various soft tissues (Sakai et al., 1990). This technique has great promise for developmental studies if young subjects are able to maintain stable postures.

Thyroid Cartilage

The newborn thyroid cartilage is broad, short, and lies close to the hyoid bone (Crelin, 1973). Bosma (1975b) considered the superior and inferior cornua of the thyroid cartilage to elongate as part of the growth process. He stated that the overall enlargement of the thyroid cartilage is more than for the cricoid. This conclusion is supported by Kahane's (1982) findings. Kahane (1982) made measurements of the thyroid cartilage from prepuberty to adulthood. Figure 18 shows the growth of the thyroid cartilage in length, height, and width. Furthermore, developmental growth patterns reveal significant sex differences. The most striking developmental sex difference is in the anteroposterior dimension. Kahane's measurements showed that the increase in the length of the thyroid cartilage is three times greater in males than in females. This difference in cartilage size is related to the growth of vocal fold length, which in males is over twice the growth present in females (Kahane, 1982). As noted earlier, the thyroid undergoes the largest increase in weight, followed by the cricoid, and then the arytenoid.

Crelin (1973) remarked that the angle between the laminae is more pronounced in males by age 3. Kahane's (1982) measurements of the angle of the thyroid laminae revealed no significant differences between males and females both at prepuberty and puberty. However, he noted significant differences in the angle of the thyroid laminae between his pubertal measures and those given in Maue (1971; cited in Kahane, 1978) for adults of both sexes. Judson and Weaver (cited in Kahane, 1978) reported the anterior angle of the thyroid cartilage to be close to 90–100° in

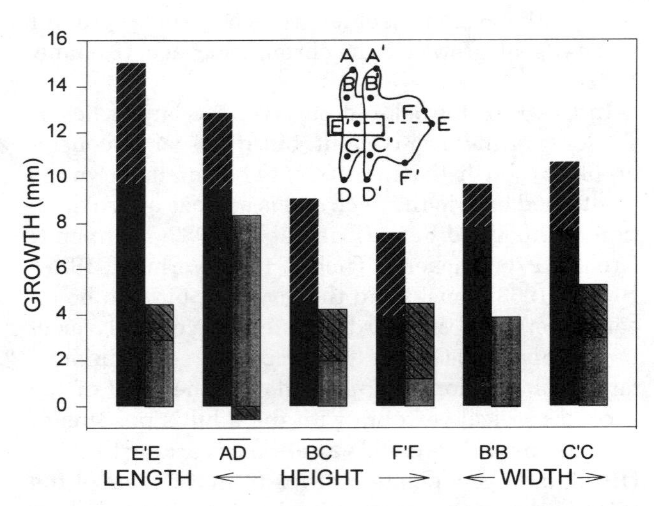

Figure 18. Growth of the thyroid cartilage expressed as increases in length, height, and width from prepuberty to puberty (black bar for males, gray bar for females) and from puberty to adulthood (diagonally slashed bars above the respective black or gray bars). Based on measurements reported by Kahane (1978, 1982). Inset drawing shows laryngeal measurement points for the thyroid cartilage.

adult males, whereas it remains relatively open at 120–130° in females. Comparable data were reported by Verhulst (1987): 90° for adult males and 120° in adult females.

The thyroid cartilage typically is cartilaginous until postadolescence. As Claassen and Kirsch (1994) note, mineralization and ossification of the thyroid cartilage begin when the larynx has attained its mature size at the age of 15 to 20 years. In contrast, the long bones exhibit these processes before attaining their final size.

Cricoid Cartilage

Tucker (1980) and Tucker and Tucker (1979) reported that in children, the cricoid arch is the most prominent midline structure in the neck below the hyoid. In contrast, in the adult, the thyroid prominence has this distinction. Tucker and Tucker (1979) described the inferior margin of the cricoid cartilage as descending during prenatal life from the second to the fourth cervical vertebra. It descends to the fifth or sixth cervical vertebra in childhood and to the level of the seventh cervical vertebra in adulthood. Fried et al. (1982) described the infant cricoid as situated at the level of the third cervical vertebra. In the adult it is at the level of the fifth vertebra. Verhulst (1987) considered the infant cricoid to be located at the level of the third cervical vertebra and then to descend progressively (level not reported).

The cricoid cartilage is the only complete ring in the laryngo-tracheo-bronchial tree. Kirchner (1988) states that the cricoid ring provides a solid base for the attachment of the muscles that dilate the glottis. According to Crelin (1973), the shape of the infant cricoid is the same as the adult cricoid. Tucker and Tucker (1979) reported the average diameter of the cricoid in infants to be 6 mm in its largest dimension and 5 mm in its smaller dimension. Too-Chung and Green (1974) described the child's cricoid as elliptical in shape with the coronal diameter being greater. However, both Sellars and Keen's (1990) and Kahane's (1978) measurements show longer anteroposterior than transverse dimensions of the cricoid lumen across development. No explanation of this discrepancy has been discussed except for possible differences in measuring technique. Mean cricoid diameters from measurements reported by Too-Chung and Green (1974) for infants birth to 2 weeks are 5.6 mm (SD = 0.053) sagittally and 5.9 mm (SD = 0.052) coronally. Sellars and Keen (1990) reported mean diameters of the cricoid for infants of the same age are 5.4 mm (SD = 0.7) sagittally and 4.9 mm (SD = 0.8) coronally. Kahane's mean sagittal diameters (J′J in Figure 19) for prepubertal males and females were 15.08 mm (SD = 2.81) and 14.54 mm (SD = 0.58), respectively, and pubertal mean sagittal diameters were 20.86 mm (SD = 2.63) and 17.23 mm (SD = 0.68), respectively. His mean coronal diameters (H′H) for prepubertal males and females were 14.18 mm (SD = 1.56) and 13.56 mm (SD = 0.86), respectively, and pubertal mean coronal diameters were 20.28 mm (SD = 2.6) and 17.23 mm (SD = 0.68), respectively.

Kahane (1982) determined that increases in the length, height, and width of the cricoid cartilage are two to three times greater in males than in females. As can be seen in Figure 19, the cricoid increases in all dimensions with posterior height increases exceeding anterior height increases. Too-Chung and Green (1974) reported the rate of growth of the cricoid to be predicted by the weight of the subject, whereas Kahane (1982) concluded that crown-heel length was the best predictor. It is noteworthy that Sellars and Keen (1990) pointed to possible racial differences in the size of the laryngeal cartilages. They observed smaller sizes in their 21 African or mixed-race cadaveric larynges aged 2 weeks to 6 months than the values reported by Too-Chung and Green (1974) for 19 Caucasian specimens from England of comparable age.

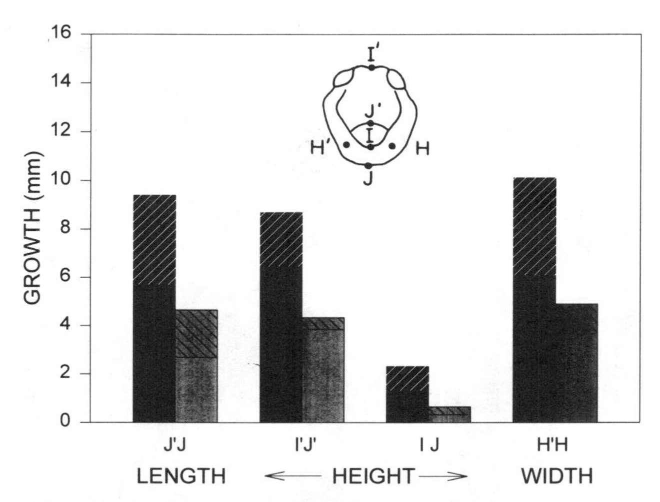

Figure 19. Growth of the cricoid cartilage expressed as increases in length, height, and width from prepuberty to puberty (black bar for males, gray bar for females) and puberty to adulthood (diagonally slashed bars above the respective black or gray bars). Based on measurements by Kahane (1978, 1982). Inset drawing shows measurement points for the cricoid cartilage.

Arytenoid Cartilage

According to Bosma (1975b), the arytenoid cartilages are relatively large at birth, so that their subsequent growth is less than for the thyroid or cricoid cartilages. As mentioned earlier in the discussion of the cricoid, Kahane's (1982) measurements of absolute weight increases of laryngeal cartilages confirmed Bosma's conclusion.

Gedgowd (1900), as reported by Kahane (1982), found the height of the arytenoid cartilages to continue growth to adulthood, whereas the width matured by age 15. Earlier work by Kahane (1978) showed that pubertal arytenoid cartilage width was not significantly different from adult widths. However, height measurements, in both sexes, differed from adult measurements. Postpubertally, the anterior edge height (measure QR in Figure 20) underwent significant growth.

As noted earlier, the arytenoid is the only laryngeal cartilage composed of both hyaline and elastic cartilage. Sato et al. (1990) stated that there is no general agreement on the distribution of elastic and hyaline cartilages in the arytenoid. They noted that some researchers have described the tip of the arytenoid to be elastic, whereas others have characterized the apex and vocal processes to be elastic (e.g., Kahane, 1983), and still others have depicted the lower part of the arytenoids to be composed of hyaline cartilage. In a

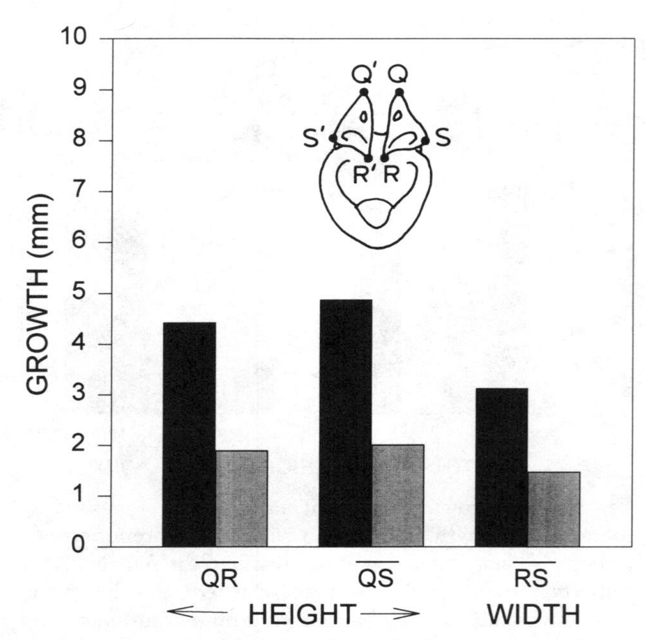

Figure 20. Growth of the arytenoid cartilage expressed as increases in height and width from prepuberty to puberty. Black bars for males, gray bars for females. Based on measurements reported by Kahane (1978). Used with permission.

study of the distribution of elastic cartilages in the arytenoids, Sato et al. (1990) showed elastic cartilage at the tip of the vocal processes as well as in the superior portion of the arytenoid, from the tip of the vocal processes to the apex, with a gradual transition from elastic cartilage to hyaline cartilage. They furthermore confirmed the presence of the elastic cartilage at birth. Sato et al. believe that the elastic cartilage serves an important role in the physiologic functioning of the arytenoid. Specifically, the elastic cartilages of the vocal processes may bend during adduction and abduction, and the arytenoids come in contact primarily at the elastic portion (i.e., superiorly). This function is maintained throughout development because elastic cartilage, as noted earlier, never ossifies.

Vocal Folds and Glottis

GROSS ANATOMIC POSITION. Because the larynx descends during development, the vocal folds have a lower relative position in adults than in infants. Verhulst (1987) described the changing position of the vocal folds relative to the cervical vertebrae. At age 5, the folds are at the level of the body of the fourth or fifth cervical vertebra. They then gradually descend to reach the inferior end of the fifth cervical vertebra, where they remain from 15–20 years of age and probably beyond.

COMPOSITION. Hirano, Kurita, and Nakashima (1981) and Hirano and Sato (1993) provided detailed descriptions of the vocal folds. They stated that, histologically, the vocal fold consists of the vocalis muscle and mucosa. The mucosa, which consists of the epithelium and three layers of the lamina propria, is proportionately thicker in newborns than in older individuals. At the medial border of the vocal fold, the ratio of the thickness of the mucosa to the length of the membranous vocal fold is about 1:2 in newborns and 1:10 in adults (Hirano et al., 1981; Hirano & Sato, 1993). The epithelium is a noncornified, stratified squamous tissue that can be likened to a thin, stiff capsule (Abitbol, 1995). Surrounding tissues of the larynx, particularly the false vocal folds and ventricles, have a covering of pseudo-stratified, ciliated columnar epithelium (Abitbol, 1995; Kahane, 1988). Structurally, the entire lamina propria in the newborn is uniform. With development, three layers evolve: a superficial layer made up of loose flexible fibers, an intermediate layer consisting primarily of elastic fibers, and a deep layer composed primarily of collagenous fibers. Hirano et al. (1981) reported that, between the ages of 1 to 4 years, the vocal ligament (intermediate and deep layers of the lamina propria) appears. Between the ages of 6 to 15 years, differentiation occurs between these two layers, such that the amount of elastic fibers and collagen increases progressively and becomes oriented along the axis of the vocal folds (deep layer of the vocal ligament). By age 16, adult morphology is present. According to Kahane (1988), the lamina propria of the adult vocal folds is approximately 1.05 mm in thickness

The histologic structure of the vocal folds continues to change with age. In the elderly, the superficial layer of the lamina propria thickens and becomes edematous, and the density of the fibers decreases. The intermediate layer becomes thinner as the elastic fibers atrophy and decrease in number. The deep layer thickens in males, especially after age 50, as the collagenous fibers increase in size and density. According to Soustin (1969), the maximal epithelial thickness in adulthood is about 180 microns in nonsmokers and perhaps twice that value in smokers. Kahane (1988) described the eight layers of stratified squamous epithelium of the vocal folds to be 0.05 mm in thickness.

Bonnaure-Mallet and Lescoat (1989) studied the distribution and development of elastic-system fibers in the vocal folds. They reported that, early in development, the elastic system consists of little amorphous material and numerous microfibrils. With development, the number of microfibrils decreases and the amount of amorphous components increases, resulting in increased elastic properties of the vocal folds. In addition to developmental changes in the mucosa, more specifically the lamina propria, Hirano, Kurita, and Nakashima (1983) commented on preliminary histologic findings concerning changes in the vocalis muscle. They observed the muscle fibers to be thin in newborns, fully developed by age 27, and atrophied with advanced age (age 61). Observations of oxytalen fibers and elastic fibers in the true vocal fold were made by Strocchi et al. (1992). Oxytalen fibers tended to be most abundant where tissue deformation was greatest, whereas more elastic fibers were found in the deepest layer of vocal ligament where the elastic constant is greatest. As Abitbol (1995) noted, the tissues of the lamina propria contain no glands, no lymphatic nodules, and relatively few blood vessels. The vocal folds depend for their humidification on glands situated in the ventricle, false vocal fold, and lower larynx.

LENGTH. Measurements of the infant vocal fold length vary somewhat among published studies (Figure 21). Some reported measurements are: 4 to 5 mm (Crelin, 1973), 4.6 mm (Kazarian, Sarkissian, & Isaakian, 1978), 6 to 8 mm (Tucker & Tucker, 1979), and 2.5 to 3 mm (Hirano et al., 1981). The measurements of Hirano et al. (1981) were based on Japanese subjects, and these measurements are the smallest. In contrast, Tucker and Tucker's (1979) measurements are the largest and apparently are based on American subjects. The discrepancies may reflect racial differences (or crown-heel length differences). Negus (1962) listed the following values of vocal fold length at different ages: 3 days, 3 mm; 14 days, 4 mm; 2 months, 5 mm; 9 months, 5.2 mm; 1 year, 5.5 mm; 5 years, 7.5 mm; 6½ years, 8 mm; and 15 years, 9.5 mm. The measurement at 3 days agrees with the value reported by Hirano et al. but is decidedly shorter than the measurements by other investigators.

Verhulst (1987) estimated the length of the vocal folds of a 6-year-old to be 8 mm. At puberty, the vocal folds lengthen to 12–17 mm in females and 15–25 mm in males. Kahane's (1982) observations reveal developmental changes from prepuberty to puberty, where the male vocal folds increase in length twice as much as the female vocal folds. Kahane reported the average adult male vocal fold length to be 28.92 mm and the female vocal fold length to be 21.47 mm. These measurements roughly correspond to measurements reported by Kazarian et al. (1978). As seen in Figure 22, their (1978) findings reflect a steady increase in vocal fold length from birth to age 16, with sex differences appearing in the age range of 6–10. Kazarian et al.'s findings also reflect a slow increase in vocal fold length, especially in males dur-

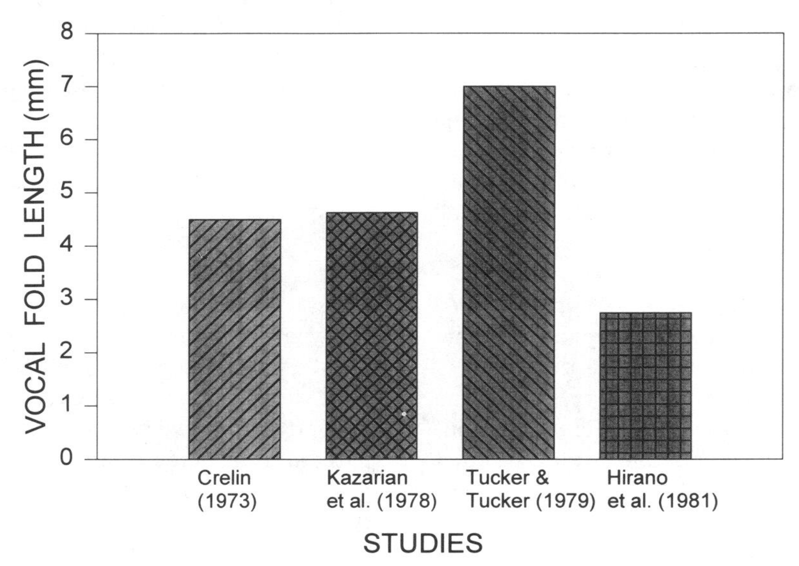

Figure 21. Measurements of vocal fold length in infants, from various published studies.

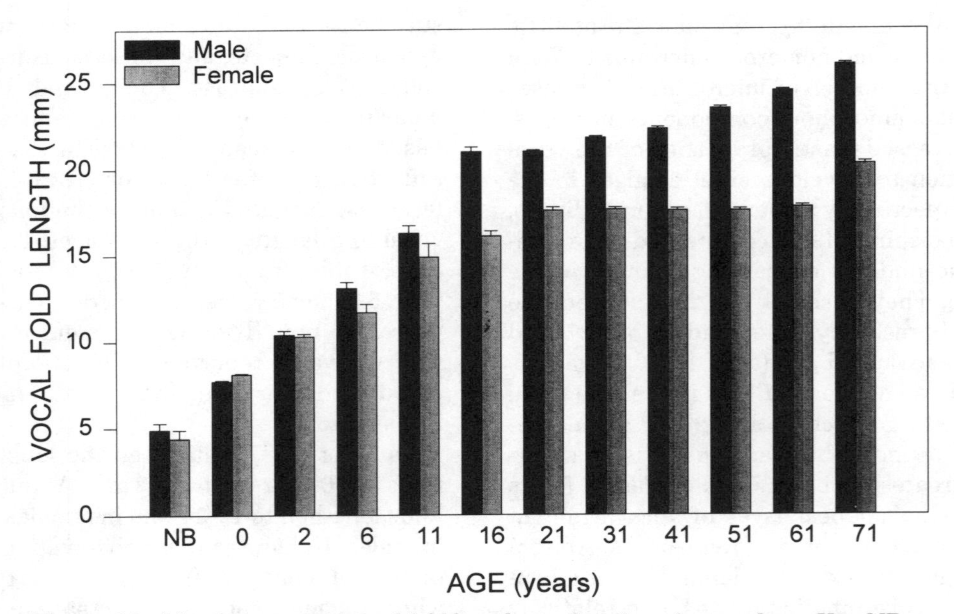

Figure 22. Measurements of vocal fold length as a function of age and sex. *Key:* NB = newborn; with the exception of NB, each age on the x-axis represents an age group spanning from the specified age to 1 month less than the following specified age (the last age group spans 71 to 90 years). Drawn from data reported by Kazarian, Sarkissian, and Isaakian (1973).

ing adulthood up to age 90. Hirano et al. (1981) did not confirm the length increases in adults, even though their study covered the age range of birth to 69 years. Kahane (1982) cited work by Gedgowd in 1900 showing that the male vocal folds continue to increase in length throughout adulthood, whereas the female vocal folds reach adult size by age 16. These length differences related to age and sex are relevant to differences in fundamental frequency, but Kahane (1982) cautioned that studies examining differences in vocal fundamental frequency between males and females should consider vocal fold thickness as well as length.

MEMBRANOUS-CARTILAGINOUS PROPORTION AND THE GLOTTIS. Despite differences in measurements used, published reports agree that the thickness of the vocal folds and the membranous portion of the glottis (space between the vocal folds) is proportionately larger in the infant than the adult. Verhulst (1987) observed that at birth the vocal fold is very wide with respect to its length. Because of the relatively large arytenoid size, one half of the infant glottal length is cartilagenous, whereas in the adult only one third is cartilagenous, with the remainder being membranous and more pliable (Tucker & Tucker, 1979).

Laitman and Crelin (1976) describe the newborn subglottal cavity as extending posteriorly and inferiorly whereas in the adult this cavity is almost vertical. The glottis of the newborn has an anteroposterior dimension of 7 mm and a posterotransverse dimension of 4 mm. The subglottic anteroposterior diameter is 5 to 7 mm at birth (Tucker, 1932, cited in Tucker, 1980). In the adult, the subglottal anteroposterior diameter approximates the maximum dimension of the glottic space.

Summary

Between birth and young adulthood, the larynx triples in size while preserving the relative weights of its intrinsic muscles; descends from the level of the second or third vertebra to the level of the fourth or fifth; and becomes firmer and less pliable. The vocal folds grow from a length of about 4 to 8 mm in the newborn to a length of about 21 mm in the adult female and about 29 mm in the adult male. The vocal folds of the newborn have a generally uniform composition and lack the distinctive layered structure found in adults. The adult-like composition is reached at about the age of 16 years. Consequently, the cover-body model of vocal fold vibration proposed for adults (Hirano & Kakita, 1985) would have to be modified to account for vibratory patterns in the infant and young child. A sexual dimorphism of laryngeal structures is evident by puberty and continues to be manifested not only in the size and configuration of the laryngeal components but also in the pattern of ossification in the adult. The larynx of the adult male is about three times heavier than that of the adult female and begins to ossify at least a decade earlier.

Clinical Note:

Larynx

Newborn infants do not have a cough reflex to water dropped on their vocal folds, but cough can be elicited as the water passes the larynx into the trachea (Kirchner, 1988). This laryngeal insensitivity, perhaps attributable to prenatal accommodation to amniotic fluid, persists for several days following birth.

Important sex differences have been described for vocal fold anatomy and physiology (Titze, 1989), although the developmental emergence of these differences needs further study. Certainly, male-female differences should be evident during puberty, if not shortly before. Hirano et al. (1981) proposed that the development of collagenous and elastic fibers of the vocal ligaments would add mass to the vocal folds and change the physical properties (e.g., stiffness) of the folds, possibly leading to voice quality changes during adolescence. According to data from Hollien, Green, and Massey (1994), the adolescent voice change in males typically begins between the ages of 12.5 and 14.5 years and has a duration of approximately 1.5 years. These authors considered adolescent voice change to be a "robust predictor of pubescence" (p. 2653). The age range of adolescent voice change described by Hollien et al. relates closely with the age range of increased vocal fold length reported by Kazarian et al. (1978).

SUMMARY OF ANATOMIC DEVELOPMENT OF THE CRANIOFACIAL-ORAL-LARYNGEAL SYSTEMS

The following is a brief summary of general issues to be considered in the assessment of speech at various stages of development.

Birth to 1 Year

The vocal tract is drastically remodeled, especially during the first 6 months of life. The oral and perioral structures initially are adapted to the requirements of suckling but begin to take on characteristics that foreshadow the vocal tract anatomy of the adult. The developmental configuration of the vocal tract in the infant is a gently curved arc, in contrast to the adult vocal tract cavity which is sharply bent around the oropharyngeal conduit (making a right-angle configuration at the craniovertebral junction). Gopal and Gerber (1992) concluded, from the configuration of the infant vocal tract, that the infant cry should be primarily vocalic and nasal and that aperiodicity or hyponasality could indicate problems such as lack of normal control, narrowing of the airway, or nasal passage obstruction. Identity of individual infants can be established with reasonable accuracy on the basis of the infant's cry (Gustafson, Green, & Cleland, 1994), which may reflect individual differences in neural control or structural properties. The infantile larynx is situated in a relatively superior position, approximately at the level of the second to third cervical vertebrae. The weights of the intrinsic muscles of the infantile larynx are proportionate to their values in the adult, with the cricothyroid muscle being the largest.

1–2 Years

The lips continue to change markedly during this interval, so that their shape is tranformed from the infantile pattern to one that conforms more to the adult shape. Relatively rapid growth also occurs in the mandible, tongue, and soft palate.

3–6 Years

Rapid cranial growth continues, with nearly maximal size reached at or shortly after 6 years. Hypertrophy of the nasopharyngeal tonsil may obstruct the nasal portal, with effects on nasal resonance in speech and on nasal resistance measured aerodynamically. By the age of about 4 years, the tongue should have extended sufficiently inferiorly so that a laryngopharynx is well developed. That is, the child's vocal tract now strongly resembles that of the adult. The vocal ligament should be evident by the age of 1–4 years, but the lamina propria of the vocal folds is still in an immature form.

6–9 Years

The intermediate and deep layers of the vocal fold lamina propria may be distinguished as early as 6 years of age, but definition of these structures may progress up to age 15 or so. A growth spurt of the mandible may occur in some individuals.

9–11 Years

Significant growth of the tongue may occur. Nasal resistance may decrease at about 9 years, presumably reflecting atrophy of lymphoid tissue.

11–18 Years

Accelerated growth of the tongue between 11 and 16 years in both sexes and accelerated growth of the mandible and lips in males can be expected between 11–16 years. During this period, the larynx reaches nearly its full size, laryngeal tissues take mature form, and the hyoid bone assumes its mature position. However, it is possible that the length of the vocal folds continues to increase in males.

ASPECTS OF MUSCLE DEVELOPMENT

The developmental picture provided by this summary is obviously incomplete; much more information is needed on several aspects. It would be very useful to have additional knowledge about the development of soft tissues such as muscle. Studies of limb muscles have shown that sex differences in muscle-fiber diameter are present by the age of 10 years (Brooke & Engel, 1969), that muscles increase in size and strength with age (Malina & Bouchard, 1991; Parker, Round, Sacco, & Jones, 1990), and that muscle relaxation times are twice as long for 3-year-olds as for 10-year-olds (Lin, Brown, & Walsh, 1994).

It is important to know if similar patterns of development apply to the muscles of speech and voice production. Compared to limb muscles, facial muscles such as the platysma, zygomaticus major, and orbicularis oris have smaller myofiber diameters, fewer Type IIB fibers, and relatively more Type II fibers not subtyped (Schwarting, Schroder, Stennert, & Goebel, 1982). Generally, facial muscles have relatively large concentrations of small-diameter myofibers and a relatively large population of Type II fibers, and these properties are linked to phasic functions in voluntary movements (Polgar, Johnson, Weightman, & Appleton, 1973; Schwarting et al., 1982; Vignon, Pellissier, & Serratrice, 1980).

Lin et al. (1994) hypothesized that motor delays in activities such as speech might be explained by aspects of muscle maturation, although the literature on development has emphasized maturation of the nervous system as the underlying factor. The longer muscle relaxation times in children may be a factor in explaining the slower speaking rates observed for young children (Haselager, Slis & Rietveld, 1991; Kent & Forner, 1980; Kent, Kent & Rosenbek, 1987; Kowal, O'Connell & Sabin, 1975). However, there is reason to suspect different maturational patterns in facial and limb muscles (Schwarting et al., 1982), so the role of muscle relaxation times in the development of speech production is uncertain. Several other aspects of muscle development could also contribute to functional differences, including rate of muscle fiber development (Sato & Sato, 1992), maturational changes in the relative numbers of Type I and Type II fibers (Kersig, cited in Verhulst, 1967), and patterns of innervation and vascularization (Fuentes & Sanchiz, 1992).

CONCLUSION

The anatomic information presented in this review is intended as a general summary of the development of the craniofacial, oral, and laryngeal systems from birth to young adulthood. Some structures exhibit changes in both shape and size, whereas others change primarily in size. The various structures that comprise the vocal tract have different periods of growth acceleration, although some of them seem to be reasonably synchronized. Because different studies focused on different structures, it is difficult to establish with certainty the degree of synchrony in growth patterns. However, a strong suggestion of coordinated development among some structures emerges from a review of the literature. A tentative conclusion is that the tongue, lips, and mandible exhibit a general synchrony of growth.

During infancy, the vocal tract structures are substantially reconfigured. The ensuing changes during childhood and adolescence may not be as dramatic in their overall effect as those observed during infancy, but alterations in size and shape continue until about 18 years in males and somewhat earlier in females. It appears that some structures (e.g., lips and vocal folds) may continue growth into adulthood, especially in males. In addition, some data indicate that the relative positions of structures may continue to change into old age. For example, laryngeal position may continue to descend during the adult years. Sexual dimorphism is a general rule of vocal tract development, with most structures exhibiting sex differences in size, shape, or both. Published studies also point to possible racial differences, but the data on this issue are quite limited.

Information on the developmental anatomy of the vocal tract should be helpful in the development of models of speech production in children and in the design of clinical evaluations and interventions for children. It appears that models of children's speech production must take account of periods of marked growth and remodeling of the vocal tract structures. Similarly, these developmental patterns should be considered in clinical protocols for the assessment of speech and other oral functions and for the effective design of clinical interventions including surgical, behavioral, and prosthetic management.

References

Adam, P., & Pohunkova, H. (1986). Ontogenesis of the elastic cartilage of the human epiglottis. A light microscopy study. *Folia Morphologica, 34,* 419–429.

Adams, M. R. (1982). Fluency, nonfluency, and stuttering in children. *Journal of Fluency Disorders, 7,* 171–185.

Allanson, J. E. (1989). Time and natural history: The changing face. *Journal of Craniofacial Genetics and Developmental Biology, 9,* 21–28.

Abitbol, J. (1995). *Atlas of laser voice surgery.* San Diego, CA: Singular Publishing Group.

Ardran, G. M., & Kemp, F. H. (1972). A functional assessment of relative tongue size. *American Journal of Roentgenology, 114,* 282–288.

Arvedson, J. C., & Rogers, B. T. (1993). Pediatric swallowing and feeding disorders. *Journal of Medical Speech-Language Pathology, 1,* 203–221.

Baer, T., Gore, J. C., Gracco, L. C., & Nye, P. W. (1991). Analysis of vocal tract shape and dimensions using magnetic resonance imaging: Vowels. *Journal of the Acoustical Society of America, 90,* 799–828.

Behlfelt, K., Linder-Aronson, S., McWilliam, J., Neander, P., & Laage-Hellman, J. (1989). Dentition in children with enlarged tonsils compared to control children. *European Journal of Orthodontics, 11,* 416–429.

Behlfelt, K., Linder-Aronson, McWilliam, J., Neander, P., & Laage-Hellman, J. (1990). Cranio-facial morphology in children with and without enlarged tonsils. *European Journal of Orthodontics, 12,* 233–243.

Behlfelt, K., Linder-Aronson, S., & Neander, P. (1990). Posture of the head, the hyoid bone, and the tongue in children with and without enlarged tonsils. *European Journal of Orthodontics, 12,* 458–467.

Behrman, R. E., & Vaughan, V. C. (1987). *Nelson textbook of pediatrics* (13th ed.). Chicago: W. B. Saunders.

Bench, R. W. (1963). Growth of the cervical vertebrae as related to tongue, face and denture behavior. *American Journal of Orthodontics, 49,* 183–214.

Bergland, O. (1963). The bony nasopharynx. *Acta Odontologica Scandinavia, 21*(Suppl. 35), 1–137.

Bernat, J. E. (1992). Pediatric dentistry. In L. Brodsky, L. Holt, & D. H. Ritter-Schmidt (Eds.), *Craniofacial anomalies: An interdisciplinary approach* (pp. 154–167). St. Louis, MO: Mosby-Year Book.

Bigenzahn, W., Piehslinger, E., & Slavicek, R. (1991). Computerized axiography for functional diagnosis of orofacial dysfunctions. *Folia Phoniatrica, 43,* 275–281.

Bishara, S. E., Peterson, L. C., & Bishara, E. C. (1984). Changes in facial dimensions and relationships between ages of 5 and 25 years. *American Journal of Orthodontia, 85,* 238–252.

Björk, A. (1955). Facial growth in man: Studies with the aid of metalic implants. *Acta Odontologia Scandinavia, 13,* 9–34.

Björk, A. (1968). The use of metallic implants in the study of facial growth in children: Method and application. *American Journal of Physical Anthropology, 29,* 243–254.

Björk, A. & Skieller, V. (1984). Growth and development of the maxillary complex. [German]. *Informationen aus Orthodontie und Kieferorthopadie mit Beitragen aus der Internationalen Literatur, 16,* 9–52.

Blocquel, H., Laude, M., Lafforgue, P., & Devillers, A. (1990). A cephalometric study of the cervical and palatal movements during growth. *Bulletin du Groupement International pour la Recherche Scientifique en Stomatologie et Odontologie, 33,* 9–18.

Bonnaure-Mallet, M., & Lescoat, D. (1989). Development of elastic system fibers in human vocal cords. *Acta Anatomica, 136,* 125–128.

Bosma, J. F. (1975a). Introduction. In J. F. Bosma & J. Showacre (Eds.), *Symposium on development of upper respiratory anatomy and function: Implications regarding sudden and unexpected infant death* (pp. 5–44). Washington, DC: Goverment Printing Office.

Bosma, J. F. (1975b). Anatomic and physiologic development of the speech apparatus. In D. B. Tower (Ed.), *The nervous system: Human communication and its disorders* (Vol. 3, pp. 469–481). New York: Raven Press.

Bosma, J. (1976). Discussion of the paper, "Postnatal development of the basicranium and vocal tract in man" by J. T. Laitman and E. S. Crelin. In J. Bosma (Ed.), *Symposium on the development of the basicranium* (pp. 219–220). DHEW Publication No. 76-989, PHS-NIH, Bethesda, MD.

Bosma, J. F. (1985). Postnatal ontogeny of performance of the pharynx, larynx, and mouth. *American Review of Respiratory Disease, 131,* S10–S15.

Broadbent, B. H., Sr., Broadbent, B. H., Jr., & Golden W. H. (1974). *Bolton standards of craniofacial developmental growth.* St. Louis: C.V. Mosby.

Brodie, A. G. (1941). On the growth pattern of the human head: From the third month to the eighth year of life. *American Journal of Anatomy, 68,* 209–262.

Brodie, A. G. (1971). Emerging concepts of facial growth. *Angle Orthodontist, 41,* 103–118.

Brooke, M. H., & Engel, K. (1969). The histographic analysis of human muscle biopsies with regard to fiber types. 4. Children's biopsies. *Neurology, 19,* 591–605.

Brulin, F., & Talmant, J. (1976). New results on lingual growth. [French]. *Revue Belge de Medecine Dentaire, 31,* 379–384.

Burke, P. H. (1980). Serial growth changes in the lips. *British Journal of Orthodontics, 7,* 17–30.

Capitanio, M. A., & Kirkpatrick, J. A. (1970). Nasopharyngeal lymphoid tissue. *Radiology, 96,* 389–391.

Castelijns, J. A., Doornbos, J., Verbeeten, B., Jr., Vielvoye, G. J., & Bloem, J. L. (1985). MR imaging of the normal larynx. *Journal of Computer Assisted Tomography, 9,* 919–925.

Castelijns, J. A., Gerritsen, G. J., Valk, J., Meyer, C. J. L. M., Kaiser, M. C., Jansen, W., & Snow, G. B. (1987). MRI of normal or cancerous laryngeal cartilages: Histopathologic correlation. *Laryngscope, 97,* 1085–1093.

Castelli, W. A., Ramirez, P. C., & Nasjleti, C. E. (1973). Linear growth study of the pharyngeal cavity. *Journal of Dental Research, 52,* 1245–1248.

Chiba, M. (1990). A comparison of growth and development of the dental arch, alveolar process, and palate in the lateral segment determined with reference to dental age and chronological age, particularly on the period of premolar eruption. [Japanese]. *Shikwa Gakuho, 90,* 909–977.

Christianson, R., Lufkin, R. B., Abemayor, E., & Hanafee, W. (1989). MRI of the mandible. *Surgical and Radiologic Anatomy, 11,* 163–169.

Claassen, H., & Kirsch, T. (1994). Temporal and spatial localization of type I and II collagens in human thyroid cartilage. *Anatomy and Embryology, 189,* 237–242.

Coccaro, P. J., & Coccaro, P. J., Jr. (1987). Dental development and the pharyngeal lymphoid tissue. *Otolaryngologic Clinics of North America, 20,* 241–257.

Cohen, S. R., Cheung, D. T., Nimni, M. E., Mahnovski, V., Lian, G., Perelman, N., & Carranza, A. P. (1992). Collagen in the developing larynx. *Annals of Otology, Rhinology and Laryngology, 101,* 328–332.

Copray, J. C. V. M., Jansen, H. W. B., & Duterloo, H. S. (1986). Growth and growth pressure of mandibular condylar and some primary cartilages of the rat in vitro. *American Journal of Orthodontics and Dentofacial Orthopedics, 90,* 19–32.

Crelin, E. S. (1973). *Functional anatomy of the newborn.* New Haven, NJ: Yale University Press.

Crelin, E. S. (1976). Development of the upper respiratory system. *Clinical Symposia, 28,* 1–30.

Croft, C. B., Shprintzen, R. J., & Rakoff, S. J. (1981). Patterns of velopharyngeal valving in normal and cleft palate subjects: A multi-view videofluoroscopic and nasoendoscopic study. *Laryngoscope, 91,* 265–271.

Croft, C. B., Shprintzen, R. J., & Ruben, R. J. (1981). Hypernasal speech following adenotonsillectomy. *Otolaryngology, Head and Neck Surgery, 89,* 179–188.

Dang, J., Honda, K., & Suzuki, H. (1993). Morphological and acoustical analysis of the nasal cavity and the paranasal cavities. *Journal of the Acoustical Society of America, 96,* 2088–2100.

Demirjian, A. (1980). Dental development: A measure of physical maturity. In F. E. Johnston, A. F. Roche, & C. Susanne (Eds.), *Human physical growth and maturation: Methodologies and factors* (pp. 83–100). NATO Advanced Study Institute Series A: Life Sciences, Vol. 30. New York: Plenum.

Diaz, G., Maccioni, P., Zedda, P., Cabitza, F., & Cortis, I. M. (1993). Dental development in Sardinian children. *Journal of Craniofacial Genetics and Developmental Biology, 13,* 109–116.

Dickson, D. R., & Maue-Dickson, W. (1982). *Anatomical and physiological aspects of speech.* Boston: Little, Brown.

Enlow, D. H. (1975). *Handbook of facial growth.* Philadelphia: W. B. Saunders.

Ettala-Ylitalo, U. M., & Laine, T. (1991). Functional disturbances of the masticatory system in relation to articulatory disorders of speech in a group of 6–8-year-old children. *Archives of Oral Biology, 36,* 189–194.

Farkas, L. G. (1992). Basic anthropometric measurements and proportions in various regions of the craniofacial complex. In L. Brodsky, L. Holt, & D. H. Ritter-Schmidt (Eds.), *Craniofacial anomalies: An interdisciplinary approach* (pp. 41–57). St. Louis, MO: Mosby-Year Book.

Farkas, L. G., Katic, M. J., Hreczko, T. A., Deutsch, C., & Munro, I. R. (1984). Anthropometric proportions in the upper-lip-lower-lip chin area in young white adults. *American Journal of Orthodontics, 86,* 52–60.

Farkas, L. G., & Posnick, J. C. (1992). Growth and development of regional units in the head and face based on anthropometric measurements. *Cleft Palate-Craniofacial Journal, 29,* 301-302.

Farkas, L. G., Posnick, J. C., & Hreczko, T. M. (1992a). Anthropometric growth study of the head. *Cleft Palate-Craniofacial Journal, 29,* 303–308.

Farkas, L. G., Posnick, J. C., & Hreczko, T. M. (1992b). Growth patterns of the face: A morphometric study. *Cleft Palate-Craniofacial Journal, 29,* 308–315.

Farkas, L. G., Posnick, J. C., Hreczko, T. M., & Pron, G. E. (1992). Growth patterns of the nasolabial region: A morphopometric study. *Cleft Palate-Craniofacial Journal, 29,* 318–324.

Fasika, O. M. (1993). Lip parameters in Nigerian children. *Plastic and Reconstructive Surgery, 91,* 446–449.

Fink, B. R., Martin, R., & LaVigne, A. B. (1975). Spring mechanics of the human larynx. In J. F. Bosma & J. Showacre (Eds.), *Symposium on development of upper respiratory anatomy and function: Implications of SID syndrome* (pp. 63–75). Bethesda, MD: U.S. Department of Health, Education, and Welfare, Publication No. (NIH) 75-941.

Fletcher, S. G. (1973). Maturation of the speech mechanism. *Folia Phoniatrica, 25,* 161–172.

Foster, T. D., Grundy, M. C., & Lavelle, C. L. (1977). A longitudinal study of dental arch growth. *American Journal of Orthodontics, 72,* 309–314.

Fried, M. P., Kelly, J. H., & Strome, M. (1982). Comparison of the adult and infant larynx. *Journal of Family Practice, 15,* 557–561.

Fuentes, A., & Sanchiz, O. (1992). Microscopic innervation and vascularization of the tongue. General study. [French]. *Revue de Stomatologie et de Chirurgie Maxillo-Faciale, 93,* 278–284.

Fujioka, M., Young, L. W., & Girdany, B. R. (1979). Radiographic evaluation of adenoidal size in children: Adenoidal-nasopharyngeal ratio. *American Journal of Roentgenology, 133,* 401–404.

Garel, C., Contencin, P., Polonovski, J. M., Hassan, M., & Narcy, P. (1992). Laryngeal ultrasonography in infants and children: A new way of investigating. Normal and pathological findings. *International Journal of Pediatric Otorhinolaryngology, 23,* 107–115.

Garel, C., Hassan, M., Legrand, I., Elmaleh, M., & Narcy, P. (1991). Laryngeal ultrasonography in infants and children: Pathological findings. *Pediatric Radiology, 21,* 164–167.

Goldstein, M. S. (1936). Changes in dimensions and form of the face and head with age. *American Journal of Physical Anthropology, 22,* 37–89.

Gopal, H. S., & Gerber, S. E. (1992). Why and how should we study infant cry? *International Journal of Pediatric Otorhinolaryngology, 24,* 145–159.

Gorlin, R. J., Pindborg, J. J., & Cohen, M. M. (1976). *Syndromes of the head and neck* (2nd ed.). New York: McGraw-Hill.

Gross, A. M., Kellum, G. D., Franz, D., Michas, K., Walker, M., Foster, M., & Bishop, F. W. (1994). A longitudinal evaluation of open mouth posture and maxillary arch width in children. *Angle Orthodontist, 64,* 419–424.

Gross, A. M., Kellum, G. D., Michas, C., Franz, D., Foster, M., Walker, M., & Bishop, F. W. (1994). Open-mouth posture and maxillary arch width in young children: A three-year evaluation. *American Journal of Orthodontics and Dentofacial Orthopedics, 106,* 635–640.

Gulisano, M., Delrio, A. N., Montella, A., Bandiera, P., & Ruggiero, C. (1992). The shape of the nasopharynx in youth: Statistical study. [Italian]. *Bollettino Societa — Italiana Biologica Sperimentale, 68,* 647–653.

Gustafson, G. E., Green, J. A., & Cleland, J. W. (1994). Robustness of individual identity in the cries of human infants. *Developmental Psychobiology, 27,* 1–10.

Hajnis, K. (1974). Kopf, Ohrmuschel-und Hand Wachstum (Verwendung bei den Operationen der angeborenen Missbildungen und Unfalls Folgen). [German]. *Acta Universitas Carolinae (Medicine), 2–4,* 77–294.

Hale, S. T., Kellum, G. D., Richardson, J. F., Messer, S. C., Gross, A. M., & Sisakun, S. (1992). Oral motor control, posturing, and myofunctional variables in 8-year-olds. *Journal of Speech and Hearing Research, 35,* 1203–1208.

Handelman, C. S., & Osborne, G. (1976). Growth of the nasopharynx and adenoid development from one to eighteen years. *Angle Orthodontist, 46,* 243–259.

Harris, J. E. (1962). A cephalometric analysis of mandibular growth rate. *American Journal of Orthodontia, 48,* 161–174.

Haselager, G. J. T., Slis, I. H., & Rietveld, A. C. M. (1991). An alternative method of studying the development of speech rate. *Clinical Linguistics and Phonetics, 5,* 53–63.

Hellman, M. (1932). An introduction to growth of the human face from infancy to adulthood. *International Journal of Orthodontics Oral Surgery and Radiography, 18,* 777.

Hirano, M., & Kakita, Y. (1985). Cover-body theory of vocal fold vibration. In R. G. Daniloff (Ed.), *Speech science* (pp. 1–46). San Diego, CA: College-Hill Press.

Hirano, M., Kurita, S., & Nakashima, T. (1983). Growth, development and aging of human vocal folds. In D. M. Bless & J. H. Abbs (Eds.), *Vocal fold physiology: Contemporary research and clinical issues* (pp. 22–43). San Diego, CA: College-Hill Press.

Hirano, M., & Sato, K. (1993). *Histological color atlas of the human larynx.* San Diego, CA: Singular Publishing Group.

Hodge, M. M. (1991). Assessing early speech motor function. *Clinics in Communicative Disorders, 1,* 69–86.

Holibkova, A. (1973). Development of laryngeal lymphatic tissue in man. *Folia Morphologica, 21,* 408–410.

Hollien, H., Green, R., & Massey, K. (1994). Longitudinal research on adolescent voice change in males. *Journal of the Acoustical Society of America, 96,* 2646–2654.

Hultcrantz, E., Larson, M., Hellquist, R., Ahlquist-Rastad, J., Svanholm, H., & Jakobsson, O. P. (1991). The influence of tonsillar obstruction and tonsillectomy on facial growth and dental arch morphology. *International Journal of Pediatric Otorhinolaryngology, 22,* 125–134.

Iseri, H., & Solow, B. (1995). Average surface remodeling of the maxillary base and the orbital floor in female subjects from 8 to 25 years. An implant study. *American Journal of Orthodontics and Dentofacial Orthopedics, 107,* 48–57.

Jansen, C., Jennekens, F. G., Wokke, J. H., Leppink, G. J., & Wijnne, H. J. (1990). Lip-length and snout indices: Methods for quantitative assessment of perioral facial muscle strength. *Journal of the Neurological Sciences, 97,* 133–142.

Jeans, W. D., Fernando, D. C., Maw, A. R., & Leighton, B. C. (1981). A longitudinal study of the nasopharynx and its contents in normal children. *British Journal of Radiology, 54,* 117–121.

Johannesson, S. (1968). Roentgenologic investigation of the nasopharyngeal tonsil in children of different ages. *Acta Radiologica (diagnosis) (Stockholm), 7,* 299–304.

Kahane, J. C. (1978). A morphological study of the human prepubertal and pubertal larynx. *American Journal of Anatomy, 151,* 11-20.

Kahane, J. C. (1982). Growth of the human prepubertal and pubertal larynx. *Journal of Speech and Hearing Research, 25,* 446–455.

Kahane, J. C. (1983). A survey of age-related changes in the connective tissue of the human adult larynx. In D. Bless & J. H. Abbs (Eds.), *Vocal fold physiology: Contemporary research and clinical issues* (pp. 44–49). San Diego, CA: College-Hill Press.

Kahane, J. C. (1988). Histologic structure and properties of the human vocal folds. *Ear, Nose and Throat Journal, 67,* 322–330.

Kahane, J. C., & Folkins, J. F. (1984). *Atlas of speech and hearing anatomy.* Columbus, OH: Charles E. Merrill.

Kahane, J. C., & Kahn, A. R. (1984). Weight measurements of infant and adult intrinsic laryngeal muscles. *Folia Phoniatrica, 36,* 129–133.

Kapur, K. K., Lestrel, P. E., Garrett, N. R., & Chauncey, H. H. (1990). Use of Fourier analysis to determine age-related changes in the facial profile. *International Journal of Prosthodontics, 3,* 266–273.

Kazarian, A. G., Sarkissian, L. S., & Isaakian, D. G. (1978). Length of the human vocal cords by age. [Russian]. *Zhurnal Eksperimentalnoi I Klinicheskoi Meditsiny, 18,* 105–109. [Note: the spelling of the authors' names is consistent with the listing in MEDLINE; the original paper gives the spelling as Ghazarian, Sargissian, & Isahakian.]

Kent, R. D. (1976). Anatomical and neuromuscular maturation of the speech mechanism: Evidence from acoustic studies. *Journal of Speech and Hearing Research, 19,* 421–447.

Kent, R. D. (1994). *Reference manual for communicative sciences and disorders: Speech and language.* San Antonio, TX: Pro-Ed.

Kent, R. D., & Forner, L. L. (1980). Speech segment durations in sentence recitations by children and adults. *Journal of Phonetics, 8,* 157–168.

Kent, R. D., Kent, J. F., & Rosenbek, J. C. (1987). Maximum performance tests of speech production. *Journal of Speech and Hearing Disorders, 52,* 367–387.

Kent, R. D., & Miolo, G. (1994). Phonetic abilities in the first year of life. In P. Fletcher & B. MacWhinney (Eds.), *Handbook of child language* (pp. 303–334). London: Blackwell.

Kerr, W. J. S., Kelly, J., & Geddes, D. A. M. (1991). The areas of various surfaces in the human mouth from nine years to adulthood. *Journal of Dental Research, 70,* 1528–1530.

Kier, W. M., & Smith, K. K. (1985). Tongues, tentacles, and trunks: The biomechanics of movement in muscular-hydrostats. *Zoological Journal of the Linnean Society, 83,* 307–324.

King, E. W. (1952). A roentgenographic study of pharyngeal growth. *Angle Orthodontist, 22,* 23–37.

Kingsbury, B. F. (1915). The development of the human pharynx. *American Journal of Anatomy, 18,* 329–347.

Kirchner, J. (1988). Functional evolution of the human larynx: Variations among the vertebrates. In O. Fujimura (Ed.), *Vocal fold physiology: Voice production, mechanisms and functions* (Vol. 2, pp. 129–134). New York: Raven Press.

Kjaer, I. (1989a). Prenatal skeletal maturation of the human maxilla. *Journal of Craniofacial Genetics and Developmental Biology, 9,* 257–264.

Kjaer, I. (1989b). Human prenatal palatal closure related to skeletal maturity of the jaws. *Journal of Craniofacial Genetics and Developmental Biology, 9,* 265–270.

Kjaer, I. (1990a). Ossification of the human fetal basicranium. *Journal of Craniofacial Genetics and Developmental Biology, 10,* 29–38.

Kjaer, I. (1990b). Radiographic determination of prenatal basicranial ossification. *Journal of Craniofacial Genetics and Developmental Biology, 10,* 113–123.

Kjaer, I., Kjaer, T. W., & Graem, N. (1993). Ossification sequence of occipital bone and vertebrae in human fetuses. *Journal of Craniofacial Genetics and Developmental Biology, 13,* 83–88.

Kowal, S., O'Connell, D. C., & Sabin, E. J. (1975). Development of temporal patterning and vocal hesitations in spontaneous narratives. *Journal of Psycholinguistic Research, 4,* 195–207.

Kramer, G. J., Hoeksma, J. B., & Prahl-Andersen, B. (1992). Early palatal changes in complete and incomplete cleft lip and/or palate. *Acta Anatomica, 144,* 202–212.

Kuehn, D. P., & Kahane, J. C. (1990). Histologic study of the normal human adult soft palate. *Cleft Palate Journal, 27,* 26–34.

Laine, T., Jaroma, M., & Linnasalo, A. -L. (1987). Relationships between interincisal occlusion and articulatory components in speech. *Folia Phoniatrica, 39,* 78–86.

Laitman, J. T., & Crelin, E. S. (1976). Postnatal development of the basicranium and vocal tract region in man. In J. Bosma (Ed.), *Symposium on the development of the basicranium* (pp. 206–220). Bethesda, MD: DHEW Publication No. 76-989, PHS-NIH.

Laitman, J. T., & Reidenberg, J. S. (1993). Specializations of the human upper respiratory and upper digestive systems as seen through comparative and developmental anatomy. *Dysphagia, 8,* 318–325.

Lam, E. W. N., Hannam, A. G., & Christiansen, E. L. (1991). Estimation of tendon-plane orientation within human masseter muscle from reconstructed magnetic resonance images. *Archives of Oral Biology, 36,* 845–854.

Lang, J., & Baumeister, R. (1984). Postnatal development of the width and height of the palate and the palate foramina. [German]. *Anatomischer Anzeiger, 155,* 151–167.

Lauder, R., & Muhl, Z. F. (1991). Estimation of tongue volume from magnetic resonance imaging. *Angle Orthodontist, 61,* 175–184.

Lavelle, C. L. (1984). A study of mandibular shape. *British Journal of Orthodontics, 11,* 69–74.

Lavelle, C. L. (1985). A preliminary study of mandibular shape. *Journal of Craniofacial Genetics and Developmental Biology, 5,* 159–165.

Lavelle C. L., & Greenwood, R. (1985). The shape of the mandible. *International Journal of Oral Surgery, 14,* 517–525.

Lee, S. K., Kim, Y. S., Lim, C. Y., & Chi, J. G. (1992). Prenatal growth patterns of the human maxilla. *Acta Anatomica, 145,* 1–10.

Lestrel, P. E. (1989). Some approaches toward the mathematical modeling of the craniofacial complex. *Journal of Craniofacial Genetics and Developmental Biology, 9,* 77–91.

Lewis, A. B. (1991). Comparisons between dental and skeletal ages. *Angle Orthodontist, 61,* 87–92.

Lieberman, P. (1984). *The biology and evolution of lanugage.* Cambridge, MA: Harvard University Press.

Lin, J. -P., Brown, J. K., & Walsh, E. G. (1994). Physiological maturation of muscles in childhood. *The Lancet, 343,* 1386–1389.

Linder-Aronson, S., & Leighton, B. C. (1983). A longitudinal study of the development of the posterior nasopharyngeal wall between 3 and 16 years of age. *European Journal of Orthodontics, 5,* 47–58.

Liversidge, H. M., Dean, M. C., & Molleson, T. I. (1993). Increasing human tooth length between birth and 5.4 years. *American Journal of Physical Anthropology, 90,* 307–313.

Lotz, W. K., D'Antonio, L. L., Chait, D. H., & Netsell, R. W. (1993). Successful nasoendoscopic and aerodynamic examinations of children with speech/voice disorders. *International Journal of Pediatric Otorhinolaryngology, 26,* 165–172.

Love, R., & Webb, R. (1992). *Neurology for the speech-language pathologist* (2nd ed.). London: Butterworth-Heinemann.

Lufkin, R. B., & Hanafee, W. N. (1989). *Pocket atlas of head and neck MRI anatomy.* New York: Raven Press.

Lufkin, R. B., Larsson, S., & Hanafee, W. (1983). NMR anatomy of the larynx and tongue base. *Radiology, 148,* 173–175.

Madzharov, M. M., & Madzharova, L. M. (1992). Age-dependent changes in the size of the upper lip in Bulgarians. *Acta Chirurgiae Plasticae, 34,* 71–78.

Malina, R. M., & Bouchard, C. (1991). *Growth maturation and physical activity* (pp. 115–132). Champaign, IL: Human Kinetics.

Mamandras, A. H. (1984). Growth of lips in two dimensions: A serial cephalometric study. *American Journal of Orthodontics, 86,* 61–66.

Mamandras, A. H. (1988). Linear changes of the maxillary and mandibular lips. *American Journal of Orthodontics and Dentofacial Orthopedics, 94,* 405–410.

Mason, R., & Simon, C. (1977). An orofacial examination checklist. *Language, Speech, and Hearing Services in Schools, 8,* 155–164.

Mazaheri, M., Krogman, W. M., Harding, R. L., Millard R. T., & Mehta, S. (1977). Longitudinal analysis of growth of the soft palate and nasopharynx from six months to six years. *Cleft Palate Journal, 14,* 52–62.

McKearn, T. W., & Stewart, T. D. (1957). *Skeletal age changes in young American males.* Natick, MA: Quartermaster Research and Development Center. Cited by Laitman and Crelin (1976).

McKerns, D., & Bzoch, K. (1970). Variations in velopharyngeal valving: The factor of sex. *Cleft Palate Journal, 7,* 652–662.

McMinn, R. M. H., Hutchings, R. T., & Logan, B. M. (1981). *A colour atlas of head and neck anatomy.* London: Wolfe Medical Publications.

Mehes, K. (1987). Head measurements in newborn infants. *Journal of Craniofacial Genetics and Developmental Biology, 7,* 295–299.

Melsen, B. (1975). Palatal growth studied on human autopsy material. A histologic microradiographic study. *American Journal of Orthodontics, 68,* 42–54.

Melsen, B., & Melsen, F. (1982). The postnatal development of the palatomaxillary region studied on human autopsy material. *American Journal of Orthodontics, 82,* 329–342.

Meredith, H. V. (1961). Serial study of change in a mandibular dimension during childhood and adolescence. *Growth, 25,* 229–242.

Merlob, P., Sivan, Y., & Reisner, S. H. (1984). Anthropometric measurements of the newborn infant (27 to 41 gestational weeks). *Birth Defects, 20*(7) 1–52.

Moore, C. A. (1992). The correspondence of vocal tract resonance with volumes obtained from magnetic resonance images. *Journal of Speech and Hearing Research, 35,* 1009–1023.

Moore, K. L. (1988). *The developing human: Clinically oriented embryology* (4th ed.). Philadelphia: W. B. Saunders.

Morris, S. E. (1982). *Pre-speech assessment scale: A rating scale for the measurement of pre-speech behaviors from birth through two years.* Clifton, NJ: J. A. Preston.

Morris, S. E., & Klein, M. D. (1987). *Pre-feeding skills.* Tucson, AZ: Communication Skill Builders.

Moss, M. L. (1968). The primacy of functional matrices on orofacial growth. *Dental Practice, 19,* 65–73.

Moss, M. L., Moss-Salentijn, L., & Ostreicher, H. P. (1974). The logarithmic properties of active and passive mandibular growth. *American Journal of Orthodontics, 66,* 645–664.

Murray, G. S., Johnsen, D. C., & Weissman, B. M. (1987). Hearing and neurologic impairment: Insult timing indicated by primary tooth enamel defects. *Ear and Hearing, 8,* 68–73.

Nanda, R. S., Meng, H., Kapila, S., & Goorhuis, J. (1990). Growth changes in the soft tissue facial profile. *Angle Orthodontist, 60,* 177–190.

Nayeem, F. (1992). The mandible — An analysis. *Acta Anatomica, 145,* 132–137.

Negus, V. E. (1962). *The comparative anatomy and physiology of the larynx.* New York: Hafner.

Nickel, J. C., McLachlan, K. R., & Smith, D. M. (1988). Eminence development of the postnatal temporomandibular joint. *Journal of Dental Research, 67,* 896–903.

Niswonger, M. E. (1934). The rest position of the mandible and the centric relation. *Journal of the American Dental Association, 21,* 1572–1581.

Omotade, O. O. (1990). Facial measurements in the newborn (toward syndrome delineation). *Journal of Medical Genetics, 27,* 358–362.

Pahkala, R. (1994). Changes in function of the masticatory system from 7 to 10 years of age in relation to articulatory speech disorders. *Journal of Oral Rehabilitation, 21,* 323–335.

Pahkala, R., Laine, T., & Lammi, S. (1991). Developmental stage of the dentition and speech sound production in a series of first-grade school children. *Journal of Craniofacial Genetics and Developmental Biology, 11,* 170–175.

Parker, D. F., Round, J. M., Sacco, P., & Jones, D. A. (1990). A cross-sectional survey of upper and lower limb strengths in boys and girls during childhood and adolescence. *Annals of Human Biology, 17,* 199–211.

Pierce, R. H., Mainen, M. W., & Bosma, J. F. (1978). *The cranium of the newborn infant.* Bethesda, MD: U.S. Department of Health, Education, and Welfare, Public Health Service (DHEW Publication No. NIH 78-788).

Polgar, J., Johnson, M. A., Weightman, D. & Appleton, D. (1973). Data on the distribution of fiber types in thirty-six human muscles: An autopsy study. *Journal of Neurological Sciences, 19,* 307–318.

Powell, T. V., & Brodie, A. G. (1964). Closure of the spheno-occipital synchondrosis. *Anatomical Record, 147,* 15–23.

Pruzansky, S. (1975). Roentgenocephalometric studies of tonsils and adenoids in normal and pathologic states. *Annals of Otology, Rhinology, and Laryngology, 84*(Suppl. 1), 55–62.

Pullinger, A. G., Liu, S. -P., Low, G., & Tay, D. (1987). Differences between sexes in maximum jaw opening when corrected to body size. *Archives of Oral Rehabilitation, 14,* 291–299.

Qvarnstrom, M. J., Jaroma, S. M., & Laine, M. T. (1993). Accuracy of articulatory movements of speech in a group of first-graders. *Folia Phoniatica, 45,* 214–222.

Qvarnstrom, M. J., Jaroma, S. M., & Laine, M. T. (1994). Changes in the peripheral speech mechanism of children from the age of 7 to 10 years. *Folia Phoniatrica et Logopedica, 46,* 193–202.

Ranly, D. M. (1988). *A synopsis of craniofacial growth.* Norwalk, CT: Appleton & Lange.

Richtsmeier, J. T., & Cheverud, J. M. (1986). Finite element scaling analysis of human craniofacial growth. *Journal of Craniofacial Genetics and Developmental Biology, 6,* 289–323.

Riolo, M. L., Moyers, R. E., McNamara, J. A., Jr., & Hunter, W. S. (1974). *An atlas of craniofacial growth.* Monograph No. 2. Ann Arbor, MI: University of Michigan.

Robbins, J., & Klee, T. (1987). Clinical assessment of oropharyngeal motor development in young children. *Journal of Speech and Hearing Disorders, 52,* 271–277.

Roy, W. L., & Lerman, J. (1988). Laryngospasm in pediatric anesthesia. *Canadian Journal of Anesthesia, 35,* 93–98.

Saito, A., & Nishihata, S. (1981). Nasal airway resistance in children. *Rhinology, 19,* 149–154.

Sakai, F., Gamsu, G., Dillon, W. P, Lynch, D. A., & Gilbert, T. J. (1990). MR imaging of the larynx at 1.5T. *Journal of Computer Assisted Tomography, 14,* 60–71.

Salinas, C. F. (1980). An approach to an objective evaluation of the craniofacies. *Birth Defects, 16,* 47–74.

Sasaki, C. T., Levine, P. A., Laitman, J. T., & Crelin, E. S. (1977). Postnatal descent of the epiglottis in man. *Archives of Otolaryngology, 103,* 169–171.

Sato, K., Kurita, S., Hirano, M., & Kiyokawa, K. (1990). Distribution of elastic cartilage in the arytenoids and its physiologic significance. *Annals of Otology, Rhinology and Laryngology, 99,* 363–368.

Sato, M., & Sato, T. (1992). Fine structure of developed human tongue muscle. *Okajimas Folia Anatomica Japan, 69,* 115–130.

Scammon, R. E. (1930). The measurement of the body in childhood. In J. A. Harris, C. M. Jackson, D. G. Patterson, & R. E. Scammon (Eds.), *The measurement of man* (pp. 173–215). Minneapolis: University of Minnesota Press.

Scheerer, W. D., & Lammert, F. (1980). Morphology and growth of the nasopharynx from three years to maturity. *Archives of Oto-Rhino-Laryngology, 229,* 221–229.

Schwarting, S., Schroder, M., Stennert, E., & Goebel, H. H. (1982). Enzyme histochemical and histographic data on normal human facial muscles. *Oto-Rhino-Laryngology, 44,* 51–59.

Scott, J. H. (1976). *Dentofacial development and growth.* Oxford: Pergamon Press.

Sellars, I., & Keen, E. N. (1990). Laryngeal growth in infancy. *Journal of Laryngology and Otology, 104,* 622–625.

Shott, S. R., & Cunningham, M. J. (1992). Apnea and the elongated uvula. *International Journal of Pediatric Otorhinolaryngology, 24,* 183–189.

Shprintzen, R. J. (1992). Assessment of velopharyngeal function: Nasopharyngoscopy and multiview videofluoroscopy. In L. Brodsky, L. Holt, & D. H. Ritter-Schmidt (Eds.), *Craniofacial anomalies: An interdisciplinary approach* (pp. 196–207). St. Louis, MO: Mosby-Year Book.

Siebert, J. R. (1985). A morphometric study of normal and abnormal fetal to childhood tongue size. *Archives of Oral Biology, 30,* 433–440.

Siebert, J. R., & Haas, J. E. (1988). Size of the tongue in sudden infant death syndrome. In P. J. Schwartz, D. P. Southall, & M. Valdes-Dapena (Eds.), The sudden infant death syndrome. *New York Academy of Sciences, 533,* 467–468.

Siegel-Sadewitz, V. L., & Shprintzen, R. J. (1986). Changes in velopharyngeal valving with age. *International Journal of Pediatric Otorhinolaryngology, 11,* 171–182.

Skolnick, M. L., McCall, G. N., & Barnes, M. (1973). The sphincteric mechanism of velopharyngeal closure. *Cleft Palate Journal, 10,* 286–305.

Skolnick, M. L., Shprintzen, R. J., McCall, G. N., & Rakoff, S. J. (1975). Patterns of velopharyngeal closure in subjects with repaired cleft palate and normal speech: A multi-view videofluoroscopic analysis. *Cleft Palate Journal, 12,* 369–376.

Slawinski, E. G., & Dubanowicz-Kossowska, E. (1993). The assessment of hypertrophy of nasopharyngeal tonsil by acoustical methods. *International Journal of Pediatric Otorhinolaryngology, 27,* 229–244.

Smahel, Z., & Skvarilova, B. (1988). Roentgencephalometric study of cranial interrelations. *Journal of Craniofacial Genetics and Developmental Biology, 8,* 303–318.

Smith, A., Weber, C. M., Newton, J., & Denny, M. (1991). Developmental and age-related changes in reflexes of the human jaw-closing system. *Electroencephalography and Clinical Neurophysiology, 81,* 118–128.

Snyder, R. G., Schneider, L. W., Owings, C. L., Reynolds, H. M., Gollomb, D. H., & Schork, M. A. (1977). *Anthropometry of infants, children, and youths to age 18 for product safety design SP-450.* Worendale, PA: Society of Automobile Engineers, Inc.

Soustin, V. P. (1969). Histochemical characteristics of the epithelium of the true vocal cords in the individual developments of man. [Russian]. *Vestnik Oto-Rino-Laringologii, 31,* 33–36.

Speirs, R. L., & Maktabi, M. A. (1990). Tongue skills and clearance of toffee in two age-groups and in children with problems of speech articulation. *Journal of Dentistry in Children, 57,* 356–360.

Sperber, G. H., & Tobias, P. V. (1989). *Craniofacial embryology.* London: Wright.

Stathopoulos, E. T. & Sapienza, C. (1993). Respiratory and laryngeal measures of children during vocal intensity variation. *Journal of the Acoustical Society of America, 94,* 2531–2543.

Strocchi, R., De Pasquale, V., Messerotti, G., Raspanti, M., Franchi, M., & Ruggeri, A. (1992). Particular structure of the anterior third of the human true vocal cord. *Acta Anatomica, 145,* 189–194.

Subtelny, J. D. (1954). The significance of adenoid tissue in orthodontia. *Angle Orthodontia,* 24, 59–69.

Subtelny, J. D. (1957). A cephalometric study of the growth of the soft palate. *Plastic and Reconstructive Surgery, 19,* 49–62.

Talmant, J., & Brulin, F. (1976). Correlative study of the sagittal development of the face and tongue (teleradiographic data-288 cases). [French]. *Orthodontie Française, 47,* 85–94.

Tamari, K., Murakami, T., & Takahama, Y. (1991). The dimensions of the tongue in relation to its motility. *American Journal of Orthodontics and Dentofacial Orthopedics, 99,* 140–146.

Tamari, K., Shimizu, K., Ichinose, M., Nakata, S., & Takahama, Y. (1991). Relationship between tongue volume and lower dental arch sizes. *American Journal of Orthodontics and Dentofacial Orthopedics, 100,* 453–458.

Teresi, L. M., Lufkin, R. B., Vinuela, F., Dietrich, R. B., Wilson, G.H., Bentson, J. R., & Hanafee, W. N. (1987). MR imaging of the nasopharynx and floor of the middle cranial fossa. Part II. Normal anatomy. *Radiology, 164,* 811–816.

Thach, B. T. (1973). Morphologic zones of the human fetal lip margin. In J. F. Bosma (Ed.) *Fourth symposium on oral sensation and perception: Development in the fetus and infant* (pp. 96–117). Bethesda, MD: DHEW Publication NIH 73-546.

Thelen, E. (1991). Motor aspects of emergent speech: A dynamic approach. In N. A. Krasnegor, D. M. Rumbaugh, R. L. Schiefelbusch, & M. Studdert-Kennedy (Eds.), *Biological and behavioral determinants of language development* (pp. 339–362). Hillsdale, NJ: Lawrence Erlbaum.

Thomas, I. T., Hintz, R. J., & Frias, J. L. (1989). New methods for quantitative and qualitative facial studies: An overview. *Journal of Craniofacial Genetics and Developmental Biology, 9,* 107–111.

Thompson, A. E., & Hixon, T. J. (1979). Nasal air flow during normal speech production. *Cleft Palate Journal, 16,* 413–420.

Titze, I. R. (1989). Physiologic and acoustic differences between male and female voices. *Journal of the Acoustical Society of America, 85,* 1699–1707.

Too-Chung, M. A., & Green, J. R. (1974). The rate of growth of the cricoid cartilage. *Journal of Laryngology and Otology, 88,* 65–70.

Tourne, L. P. (1991). Growth of the pharynx and its physiologic implications. *American Journal of Orthodontics and Dentofacial Orthopedics, 99,* 129–139.

Tracy, W. E., & Savara, B. S. (1966). Norms of size and annual increments of five anatomical measurements of the mandible. *Archives of Oral Biology, 11,* 587–598.

Tucker, G. F., Jr. (1980). Laryngeal development and congenital lesions. *Annals of Otology, Rhinology and Laryngology, 89*(5, Pt. 2, Suppl.), 142–145.

Tucker, J. A., & Tucker, G. F. (1979). A clinical perspective on the development and anatomical aspects of the infant larynx and trachea. In G. B. Healy & T. J. I. McGill (Eds.), *Laryngo-tracheal problems in the pediatric patient* (pp. 3–8). Springfield, IL: Charles C. Thomas.

Tucker, J., Vidic, B., Tucker, G. F., Jr., & Stead, J. (1976). Survey of the development of laryngeal epithelium. *Annals of Otology, Rhinology and Laryngology, 52*(5, Suppl. 30, Pt. 2), 1–16.

Ursi, W. J., Trotman, C. A., McNamara, J. A., Jr., & Behrents, R. G. (1993). Sexual dimorphism in normal craniofacial growth. *Angle Orthodontist, 63,* 47–56.

Valadian, I., & Porter, D. (1977). *Physical growth and development: From conception to maturity.* Boston: Little, Brown.

van Spronsen, P. H., Weijs, W. A., Valk, J., Prahl-Andersen, B., & van Ginkel, F. C. (1991). Relationships between jaw muscle cross-sections and craniofacial morphology in normal adults, studied with magnetic resonance imaging. *European Journal of Orothodontics, 13,* 351–361.

Verhulst, J. (1987). Development of the larynx from birth to puberty. [French]. *Revue De Laryngologie Otologie Rhinologie, 108*(4), 269–270.

Vig, P. S., & Cohen, A. M. (1979). Vertical growth of the lips: A serial cephalometric study. *American Journal of Orthodontics, 75,* 405–415.

Vignon, C., Pellissier, J. F., & Serratrice, G. (1980). Further histochemical studies on masticatory muscles. *Journal of Neurological Science, 45,* 157–176.

Vogl, T., Bruning, R., Grevers, G., Mees, K., Bauer, M., & Lissner, J. (1988). MR imaging of the oropharynx and tongue: Comparison of plain and Gd-DTPA studies. *Journal of Computer Assisted Tomography, 12,* 427–433.

Walker, G. F., & Kowalski, C. J. (1972). On the growth of the mandible. *American Journal of Physical Anthropology, 36,* 111–118.

Ward, R. E. (1989). Facial morphology as determined by anthropometry: Keeping it simple. *Journal of Craniofacial Genetics and Developmental Biology, 9,* 45–60.

Warren, D. W., Duany, L. F., & Fischer, N. D. (1969). Nasal pathway resistance in normal and cleft lip and palate subjects. *Cleft Palate Journal, 6,* 134–140.

Webber, R. L., & Blum, H. (1979). Angular invariants in developing mandibles. *Science, 206,* 689–691.

Widmer, R. P. (1992). The normal development of teeth. *Australian Family Physician, 21,* 1251–1261.

Wind J. (1970). *On the phylogeny and ontogeny of the human larynx.* Groningen, The Netherlands: Wolters-Noordoff Publishing.

Wittmann, A. L. (1977). Macroglossia in acromegaly and hypothyroidism. *Virchows Archives of Pathological Anatomy and Histology, 373,* 353–360.

Wood, J. L., & Smith, A. (1991). Cutaneous oral-motor reflexes of children with normal and disordered speech. *Developmental Medicine and Child Neurology, 33,* 797–812.

Wortham, D. G., Hoover, L. A., Lufkin, R. B., & Fu, Y. S. (1986). Magnetic resonance imaging of the larynx: A correlation with histologic sections. *Otolaryngology, Head and Neck Surgery, 94,* 123–133.

Wowern, N. V., & Stoltze, K. (1979). Comparative bone metamorphometric analysis of mandibles and second metacarpals. *Scandinavian Journal of Dental Research, 87,* 258–364.

Wright, A., Ardan, G. M., & Stell, P. M. (1981). Does tracheostomy in children retard the growth of trachea or larynx? *Clinical Otolaryngology, 6,* 91–96.

Zemlin, W. R. (1988). *Speech and hearing science: Anatomy and physiology* (3rd ed.). Englewood Cliffs, NJ: Prentice-Hall.

Zinreich, S. J., Kennedy, D. W., Kumar, A. J., Rosenbaum, A. E., Arrington, J. A., & Johns, M. E. (1988). MR imaging of normal nasal cycle: Comparison with sinus pathology. *Journal of Computer Assisted Tomography, 12,* 1014–1019.

Zylinski, C. G., Nanda, R. S., & Kapila, S. (1992). Analysis of soft tissue facial profile in white males. *American Journal of Orthodontics and Dentofacial Orthopedics, 101,* 514–518.

Appendix

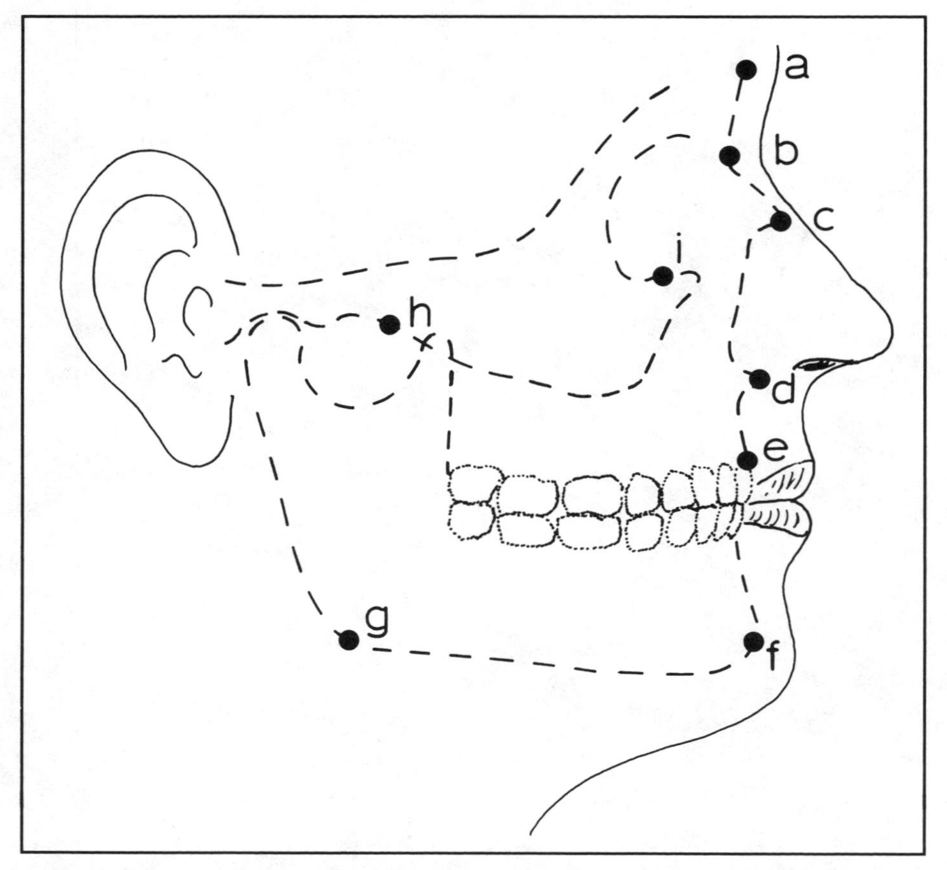

Figure A–1. Major anthropometric points of the skull and face. *Key:* **a** = glabella, **b** = nasion, **c** = rhinion, **d** = nasospinale, **e** = prosthion, **f** = mention, **g** = gonion, **h** = zygion, **i** = orbitale.

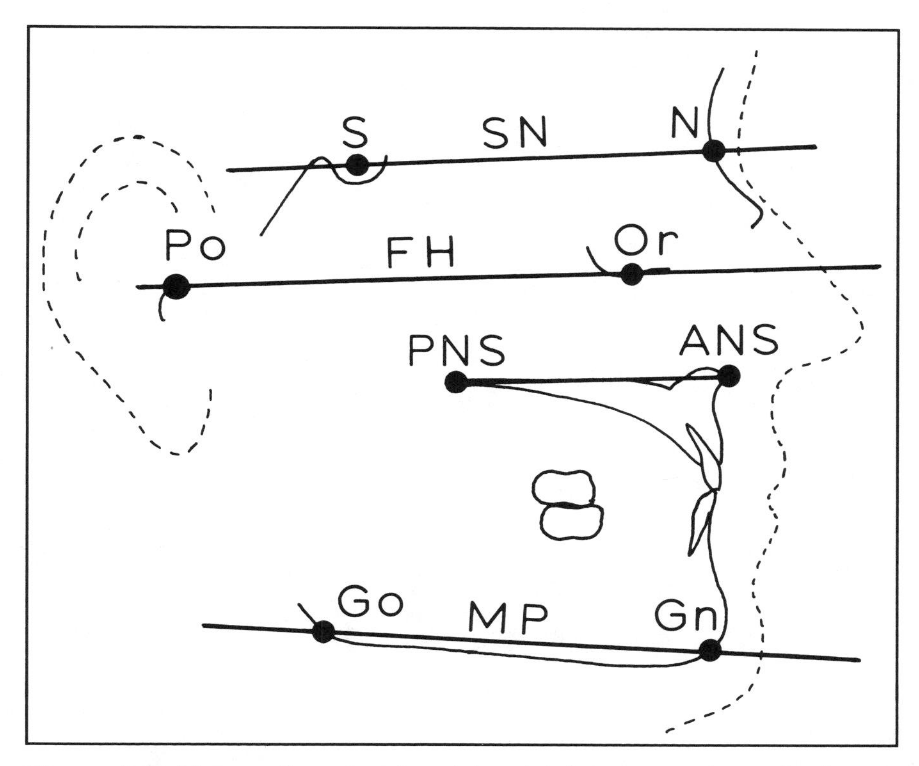

Figure A–2. Major anthropometric points related to commonly used reference lines. *Key:* **S** = sella, **N** = nasion, **SN** = sella-nasion line, **Po** = porion, **Or** = orbitale, **FH** = Frankfort line, **PNS** = posterior nasal spine, **ANS** = anterior nasal spine (PNS-ANS is the line drawn to connect PNS and ANS), **Go** = gonion, **Gn** = gnasion, **MP** = mandibular plane.

Index

O

P

S

T

V

NORTHERN MICHIGAN UNIVERSITY
3 1854 005 438 901